ORGANIZATIONAL FORM *and* BUSINESS STRATEGY *in the* U.S. PETROLEUM INDUSTRY

Michael Ollinger

UNIVERSITY
PRESS OF
AMERICA

Lanham • New York • London

338.27282
0490

Copyright © 1993 by
University Press of America®, Inc.
4720 Boston Way
Lanham, Maryland 20706

3 Henrietta Street
London WC2E 8LU England

Library of Congress Cataloging-in-Publication Data

Ollinger, Michael.
Organizational form and business strategy in the U.S. Petroleum
industry / Michael Ollinger.
p.   cm.
Includes bibliographical references and index.
1. Petroleum industry and trade—United States.   I. Title.
HD9565.O648   1993
338.2'7282'0973—dc20   92–41876 CIP

ISBN 0–8191–8990–1 (cloth : alk. paper)

The paper used in this publication meets the minimum requirements of
American National Standard for Information Sciences—Permanence
of Paper for Printed Library Materials, ANSI Z39.48–1984.

Acknowledgments

I would like to express my heartfelt thanks to Seth Norton, Gary Miller, Bob Parks, and Lee Benham of Washington University in St. Louis who gave me many valuable suggestions in the writing of this book.

Contents

List of Tables

List of Figures

Chapter 1

Introduction

A number of prominent economists (Alchian and Demsetz, 1972; Jensen and Meckling, 1976; Williamson, 1975) follow Coase (1937) in affirming that transaction costs determine organizational form. The precise nature of the economic forces driving organizational structure, however, is subject to debate. Some theorists (Williamson, 1975; Armour and Teece, 1978; Teece, 1981) believe that firm size determines organizational form. They argue that as the number of interactions between workers rises, employees lose the incentive to produce efficiently. Williamson asserts the following (1975, p. 150):

> The organization and operation of the large enterprise along the lines of the M-form (multidivisional form) favors goal pursuit and least cost behavior more nearly associated with the neoclassical profit maximizing hypothesis than does the U-form (functional alternative).

Other theorists (Chandler, 1977; Rumelt, 1986) believe that it is not size but the diversity of transactions that determines organizational structure. They claim that businesses become more difficult to manage as firms grow into different product and geographic markets. Chandler indicates the following (1977, p. 463):

> The first (functional organizational form) has been used primarily by companies producing a single line of goods for one major product or regional market, the second by those manufacturing several lines for a number of product and several lines for a number of product and regional markets

The purpose of this book is to extend the work of other theorists on the multidivisional form and to specify the conditions under which alternative organizational structures are efficient. Single-product firms are argued to be as efficiently organized as functional firms, but, after diversifying, lose their efficiency and therefore change to the multidivisional form as a way to lower the organizational costs of operating in heterogeneous product and geographic markets. Moreover, the multidivisional form is suggested to be better able to accomodate related, rather than unrelated, business activities.

Anecdotal evidence for single-product oil firms offers some support for the view that diversity, not size, determines organizational form. Kerr-McGee, for example, adopted the multidivisional form in 1954 at a size that was 15 times less than that of Gulf Oil, which changed in 1957, and 10 times less that of Sun, which changed in 1967. All three firms were in the process of diversifying when they changed their structure.[1]

Previous studies (Hill, 1985; Steer and Cable, 1978; Thompson, 1981) found in a sample of U.K. firms that multidivisional firms outperformed holding companies. Cable and Dirrenheimer (1983), however, reported that West German multidivisional firms performed no better than other firms. Teece (1981) and Harris (1983) provide ambiguous evidence from their tests of whether the adoption of the multidivisional form is crisis induced. Each compared the ex-ante performance of selected firms with their ex-post performance with Teece finding support for an increase in profitability and Harris finding no difference in performance.

Several criticisms can be made of these studies. First, all of the theorists used accounting measures of profitability and may have reported biased results.[2] Second, none of the scholars specify conditions under which alternative organizational arrangements may be optimal and cannot explain why nondivisional forms persist. Third, the studies are contradictory, except for a general finding that the holding company form is inefficient relative to the multidivisional form. Fourth, only Cable and Dirrenheimer (1983) controlled for the transition period.[3]

---

[1]Similar evidence is found for firms outside of the oil industry. Using Bhargava's (1972) organizational form data and accounting data from *Moody's Industrial Manual*, one can note that, in the chemical industry, Monsanto changed organizational form when it was four times smaller than Kodak. When Kodak changed structure, it was about 40 times larger than Monsanto at the time of its change. Similarly, in the tire industry, Goodrich changed form when it was half as big as Goodyear, a firm that did not change form until 30 years later when it was six times the size of Goodrich. Finally, in the auto industry, American Motors changed form when it was eight times smaller than Ford.

[2]Bentson (1986) and Fisher and McGowen (1983) claim that accounting rates of return are distorted by the failure to consider differences in systematic risk, temporary disequilibrium effects, tax laws, and arbitrary accounting conventions.

[3]Cable and Dirrenheimer (1983) found this period to be significantly more costly than times when the firm is alternatively organized.

In the only direct test of Williamson's hypothesis, Armour and Teece (1978) tested for profitability differences in a sample of oil firms. They found that multidivisional oil firms had higher after-tax profits than functionally organized oil firms from 1955 to 1968 but not from 1969 to 1973. Their empirical work, however, raises numerous questions. First, higher after-tax profits did not exist from 1969 to 1973. Second, economic rents stemming from low-cost sources of oil were not considered. Third, the effect of taxes on after-tax profitability may have biased the results.

Previous work indicates that most large modern corporations adopt the multidivisional form, and yet results of differences in performance among alternative organizational arrangements are ambiguous. The differences raise several questions concerning the accuracy of the existing theory. First, according to both Williamson (1975) and Armour and Teece (1978), the multidivisional form offers a control technique similar to that of the capital markets while gaining superior internal informational advantages. For centralized control, information at the central office must be identical to that at the division level. If it is not identical, then decentralized managers can manipulate costs and behave opportunistically, but if it is then the centralized office can interfere with daily decision-making. Hence, optimum control requires a fine balance which, when combined with opportunistic managers at either level, may be impossible to achieve

Second, both Williamson (1975) and Armour and Teece (1978), fail to consider decentralized functional alternatives. Yet Chandler (1977) has noted two: the decentralized General Electric form and the decentralized railroad form developed at the Pennsylvania Railroad. Chandler indicates the following (1977, p. 179):

> This [decentralized railroad] structure, with its autonomous subsystem responsible for day-to-day operations and its general office to handle long-term supervision and planning, was as sophisticated as any modern giant industrial enterprise...For Perkins (the top manager), the most important duties of the top managers in the general offices were strategic planning and recruitment of senior managerial personnel.

Third, Williamson's interpretation of Chandler's case studies of General Motors and DuPont was of firms suffering from inadequate centralized control, neither of which accurately reflected a transition from a well-managed functional firm to a multidivisional firm. Hence, it is difficult to distinguish the effect of multidivisionalization from the impact of better control.

Fourth, Williamson (1975) and Armour and Teece (1978) believe that the lag in the spread of the multidivisional form from General Motors in 1921 to most other firms between 1950 and 1970 was a simple diffusion process. If this were true, however, competitive market forces must either be very weak or the multidivisional form must be an ambiguously superior organizational form.

Based on the foregoing criticisms and the hypothesis that single-product firms are efficiently organized within a functional organizational framework, five types of empirical and historical measures are used to test the hypothesis. First, event studies of the wealth effects associated with the multidivisional form are used to evaluate economic profits. Second, cross-sectional tests for each year in the test period are used to evaluate the long-run relative profitability of alternative organizational arrangements. Third, a set of time series tests at the firm level are used to evaluate changes in long-run relative profitability, operating profitability, managerial efficiency, and changes in diversified sales for individual firms. Fourth, historical evidence from 1950 to 1973 relates adoption of the multidivisional form to diversification into related businesses. Fifth, data on business activity and historical evidence from the 1973-90 period contrasts the relative efficiencies of related versus unrelated growth strategies.

The results support four findings related to the hypothesis. First, there is no performance difference between a single-product functionally organized firm and a multiproduct multidivisional firm. Second, multiproduct decentralized organizational alternatives to the multidivisional form are inefficient. Third, functionally organized firms that are in the process of diversifying adopt the multidivisional form rather than an alternative as a low-cost way of managing multiple product lines. Fourth, the most efficient use of the multidivisional form can be achieved with related growth strategies in which similar job responsibilities can be centralized, and information from one division can be used to benefit the operations of another unit.

In this book Chapter 2 is a description of alternative organizational forms, Chapter 3 is a theory of alternative organizational arrangements, Chapter 4 presents firm histories and strategies, Chapter 5 presents empirical models, Chapter 6 contains a description of the data, Chapter 7 presents the results, Chapter 8 contains historical analysis of the 1973-90 period and pertains to the applicability of the multidivisional form for unrelated and related growth strategies, and Chapter 9 has a discussion and conclusion. The appendix contains an historical overview of the oil industry between 1945 and 1990 as well as supplementary tables.

Chapter 2

Alternative Organizational Arrangements

2.1 Overview

Corporate organizational forms can be place on a schema measuring the degree of decision-making centralization (figure 2.1).

FIGURE 2.1-ALTERNATIVE FORM SCHEMA

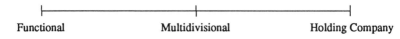

At the intermediate node lies the multidivisional form, which decentralizes decision-making for daily operations, while centralizing information used in planning and control. At the ends of the schema are the two extreme cases of firm organizational arrangements: the highly centralized functional form and its decentralized counterpart, the holding company form. Between these three general forms, variant organizational configurations given by product market and firm operating needs may also exist.

2.2 Functional Form

Specialization by managerial function permits economies in decision-making and an efficient division of labor, according to Ansoff and Brandenburg (1971). Chandler (1977) observes that as railroads in the mid-1800's grew beyond local markets, they encountered administration costs that rose more rapidly than revenues. He suggests that the early railroad managers reduced costs by first dividing management positions into line and staff management roles, further decomposing staff functions into functional groupings (such as accounting, planning, and traffic depart-

functions into functional groupings (such as accounting, planning, and traffic departments). Functional teams were then supplemented as the firm either grew or the market environment changed.

As large single-product firms emerged from the industrial sector, they adopted the centralized functional organizational model. These firms introduced coordinating committees comprised of the heads of various departments that met to expedite product flows, rate of return criteria for planning and control functions, and decentralization of departments to give managers more time for making strategic plans. According to Chandler, however, two imperfections became apparent as single-product firms diversified into multiple products. First, excessive centralization caused slow firm responsiveness to changing market conditions. Second, functional specialization led to planning by negotiation among top functional managers who were primarily interested in their own departments rather than the welfare of the firm (Chandler, 1977, pp. 453-4).

2.3 Holding Company

Williamson describes the holding company as a loosely divisionalized firm with little centralized control over separate operating units. The holding company form evolved during 1890-1905 and, a legal precedent, the E. C. Knight case, that established the New Jersey Holding Company as a legal entity (Chandler, 1977, p. 333). Prior to this case, most holding companies acted as cartels in which a central office both held the stock of subsidiaries and set price and output quotas, but had no operating control. Afterward, many holding companies centralized administration and integrated operating systems. Chandler claims those that did not centralize and integrate either were forced into bankruptcy or were dissolved.[1]

Change occurs at holding companies through the acquisition and sale of businesses. New firms are incorporated into its structure by lateral expansion, while old units are removed by horizontal contraction. In this framework, duplication of managerial effort can exist because similar horizontal units are not consolidated. Moreover, when growth occurs in the product lines of sub-units, the firm, like a functional firm, must grow radially, resulting in a slower responsiveness to market changes.

2.4 Multidivisional Form

The multidivisional form provides an intermediate degree of centralization of management control in which independent divisional sub-units have complete authority over daily decision making and the central office provides strategic planning and division control functions. According to Chandler, this form evolved during the post-World War I recession when rapid changes in demand conditions and slow

---

[1]Chandler cites the analysis of successful and unsuccessful mergers made by Shaw Livermore in 1935 in "The Success of Industrial Mergers," *Quarterly Journal of Economics*, 50 (November): 94.

inventory adjustments left many centralized firms near bankruptcy. Chandler summarized the changes at DuPont and General Motors as follows (1977, p. 457):

> In this [multidivisional] structure, autonomous divisions continued to integrate production and distribution by coordinating flows from suppliers to consumers in different, clearly defined markets. The divisions, headed by middle managers, administered their functional activities through departments organized along the lines of those at General Electric and DuPont. A general office of top managers, assisted by large financial and administrative staffs, supervised these multifunctional divisions. The general office monitored the divisions to be sure that their flows were tuned to fluctuations in demand and they had comparable policies. The top managers also evaluated the financial and market performance of the divisions. Most important of all, they concentrated on planning and allocating resources.

According to Williamson (1975), optimum control requires the firm to: (1) identify separable economic activities, (2) organize divisions as sub-units and give them semi-autonomous standing, (3) monitor division performance, (4) award incentives to managers, (5) allocate cash flow to the best performing units, and (6) provide strategic planning for the entire firm.

2.5 Hybrid Organizational Forms

Theorists (Armour and Teece, 1978; Chandler, 1977; Rumelt, 1986) have observed three hybrid organizational forms that combine features of the three primary organizational forms. Rumelt (1986) argues that the functional-with-subsidiary form (F.S.) is used by firms that seek to stress one primary marketing activity over secondary market interests. He indicates that in its centralized structure the top manager oversees a second layer of management consisting of a group of functional managers devoted to a primary market and division (subsidiary) presidents responsible for secondary market activities.

Chandler (1977) observed a centralized F.S. at Standard Oil of New Jersey. It employed a centralized coordination and budget department to process production information for use by top management, a treasurer's and comptroller's office to deal with financial reporting, and a centralized marketing and manufacturing department to coordinate activities with their counterparts at subsidiaries. The subsidiaries were established on a regional or technological basis, and organized on a functional basis with subsidiary managers making all operating decisions. Chandler, however, notes that neither foreign nor domestic subsidiaries were completely consolidated but top management did offer centralized control over a decentralized structure.

Greene (1985) and Chandler (1977) observed several problems in Mobil's decentralized F.S. First, they indicated that Mobil grew through acquisitions that were never completely consolidated. As a result, subsidiary managers promoted their interests over those of the firm. Second, subsidiaries were separated along both geographical and functional lines with no mechanism to coordinate flows between functions or across regions. Third, Chandler (1977) asserts that both accounting controls and long-run planning were weak and not modernized until after the adoption of the multidivisional form.

A second hybrid functional organizational form was noted by Chandler (1977). The decentralized railroad form arose as railroads expanded geographically, and excessive centralization placed extreme time demands on top managers. In response, railroad managers developed a more decentralized organizational form in which general managers headed semi-autonomous divisions with their own traffic, transportation, accounting, and purchasing departments. They decentralized all daily decision making but retained centralized functional control over accounting and purchasing to gain economies in those departments. Chandler suggests that the efficiencies were such that alternative organizational forms survived only because they were able to obtain favorable treatment under regulation.

Another hybrid managerial arrangement was observed by Chandler (1977) at General Electric. Rather than centralizing all daily decision making, it decentralized some responsibilities along functional lines and used coordinating committees to expedite information among functions.

Chandler (1977) also noted a final form, the centralized holding company structure. He claims that the loosely divisionalized holding company structure discussed above and by Williamson (1975) ceased to exist after 1917. Replacing it was a more centralized structure which was driven by competitive forces to centralize control over production, administration, marketing, and purchasing.

Chapter 3

Theory of Alternative Organizational Arrangements

3.1 Overview
Many economists (Silver and Auster, 1969; Lucas, 1978; Oi, 1983) view time constraints on top managers as a restriction to firm growth. In small functionally organized firms, top managers have sufficient time for both monitoring daily activities and making strategic plans. But as the firm grows, efficient control and planning may require more time than a single manager can provide. Chandler (1977) makes two relevant observations of highly centralized firms. The first concerns firm responsiveness to fluctuations in demand in the wake of the post-World War I recession (1977, pp. 456-57):

> The slowdown in demand caught both mass marketers and large integrated industrials by surprise....The large integrated manufacturers and processors in the chemical and mechanical industries, where a much longer period of time was required to get costly materials through the processes of production and distribution, had the greatest difficulty of all....[Many] manufacturers had to follow the General Motors example and partially write down the value of their overstocked inventory. General Electric, U.S. Rubber, and other large enterprises responded by developing techniques that set and adjusted their flows to carefully forecasted future demand. At General Motors and DuPont, the reorganization went further (and created the multidivisional form)...

The second relates to management attitudes (1977, p. 453):

> ...in the centralized functionally departmentalized organizations, top managers responsible for long-term allocations continued to concentrate on daily opera-

...in the centralized functionally departmentalized organizations, top managers responsible for long-term allocations continued to concentrate on daily operations... despite admonitions from Coleman and Pierre DuPont, [DuPont] managers preferred to give priority to more immediate problems...than to what seemed to them vague and less pressing concerns - long-term planning and appraisal.

Excess centralization may impede firm responsiveness and provide little time for strategic planning, but inadequate centralization can lead to a loss of firm direction and shirking of responsibilities by subordinate managers. Decentralized firms, such as holding companies, and entrepreneurial firms place most power in the hands of subordinate managers and benefit neither from internalized auditing nor from firm-wide guidance. Williamson asserts that in its highly decentralized form, holding companies can do no better than a well-diversified portfolio of firms under the direction of an investment banker. Moreover, Chandler notes that before their changes to more centralized organizational forms, decentralized control at Armour led to ad hoc growth motivated by middle managers, while it almost led to the demise of General Motors.

Scholars (Chandler, 1977; Rumelt, 1986) observed four structures used by entrepreneurs to reduce the costs of excessive or inadequate centralization: the functionally organized decentralized railroad model, the centralized holding company form, the F.S., and the multidivisional form. All four can provide centralized control and planning, but they differ by how they identify separable economic activities and their degree of decentralized control. Only the multidivisional form makes a complete departure from functional specialization. It establishes a division of labor determined by decision-making roles in which managers at the central office specialize in strategic planning and sub-unit control, while managers at the division level make all daily decisions. The hybrid organizational forms, in contrast, require that top managers make at least some of the daily decisions and that some functional managers report to the executive manager.

A change in the managerial division of labor requires a structural organizational change. Functional firms take a pyramid-shaped form with the top manager at its apex, while holding companies have a flat structure with little centralized control. The multidivisional form combines features of both. Similar to a holding company, the multidivisional form has a flat structure and, similar to the functional form, has a chief decision maker at the top of its structure. Keren and Levhari (1979) have shown that a flat structure provides a centralized firm with greater planning time and quicker responsiveness to changing market conditions but at a higher fixed cost.

The fixed cost-greater planning time trade-off prompts two questions. First, under what circumstances is the firm willing to incur higher fixed costs in order to change organizational form? Second, can alternative organizational arrangements substitute under some conditions and provide similar benefits? Within this context, contractual theory of the firm can be used to analyze the rationale for the adoption of the multidivisional form.

3.2 Functional Organizational Structure

Oi (1983) argues that firms organize production around teams to capture the gains of specialization. One way for the firm to organize managerial relations is by functional specialization. Ansoff and Brandenburg (1971) assert that this type of management structure is optimal for a stable environment in which functional managers are matched with market characteristics.

For single-product firms, change occurs in the product environment. The functionally organized firm accommodates it by adding functions and expanding laterally. According to Keren and Levhari (1979), horizontal growth provides more planning time and quicker responsiveness to market changes.

As functionally organized firms grow into new markets, organizational costs may rise. Rubin (1978) argues that agency costs rise with the distance from centralized control. Moreover, according to Williamson (1975), excessive centralization may impose extreme time demands on top managers. Accordingly, as firms grow beyond local markets, organizational change may be necessary.

If product characteristics are homogeneous in new geographic markets, then a simple adaptation of the functional form may be possible. Chandler (1977) indicates that two single-product firms mainly operating in the U. S. market, General Electric and DuPont, overcame the costs associated with geographic distances by decentralizing some daily decision making to regional offices. As a result, they were able to retain the advantages of functional specialization while being able to operate in a distant but homogeneous market.

Growth into heterogeneous markets may be more costly than expansion into new geographic regions. If the firm matches each functional manager with one market characteristic and the number of characteristics rises, then the number of functional managers must also rise. Hence, if all responsibilities are centralized, then team size must rise and eventually press the limit of the span of control, resulting in both shirking of responsibilities by subordinate managers and inadequate time for decision making by the top manager.[1]

---

[1]Keren and Levhari have shown that the span of control depends on fixed time per level and the number of exchanges between executives and subordinates. Alchian and Demsetz (1972) argue that as team size rises measurement of any one member become more imprecise and, therefore, shirking rises. Similarly, Williamson (1970) asserts that functionally organized firms grow radially (increased team size), which increases the number of exchanges between organizational levels, and results in an increase in opportunism in the system.

One way for firms to reduce the size of one functional team is to create multiple teams of functional managers, each reporting to a manager that is subordinate to the top manager. This may be efficient for small firms, but for large firms the resulting increase in the span of control can press the limits of managerial time. Further, measurement problems of subordinate managers arises as subordinate managers are evaluated on the basis of effort in multiple product markets. Moreover, Williamson (1975) argues that competition for resources among functional managers encourages them to stress functional goals rather than firm goals, raising the costs of opportunism.

3.3 Functional-with-Subsidiaries Organizational Structure

An alternative way of reducing organizational costs is to combine functional responsibilities devoted to each secondary marketing activity into separate teams of functional managers that are overseen by general managers. Under this arrangement, a top manager can monitor the general manager by matching managerial performance with product performance in the marketplace.[2] Furthermore, if firm activities in secondary markets are not significant and the firm wants to focus its attention on its primary business, then the top manager may still act as the general manager of the primary line of business. Accordingly, the firm adopts the F.S. form. Under this arrangement, the firm makes its best decision maker responsible for both daily decisions in the primary market and strategic planning for all of its businesses.[3]

For FS-organized firms, change comes either from a new product environment of the primary market or through new business opportunities. If the product environment of the primary market changes, then the firm, like a functional firm, can grow laterally in new product responsibilities; if new businesses are added, similar to a multidivisional firm, the firm can grow laterally in sub-units. According to Keren and Levhari (1979), however, there is a limit to the span of control of any executive and, hence, a limit to the amount of efficient horizontal growth. If, for example, the firm decides to place emphasis on its secondary markets, there may be insufficient time for the top manager to monitor daily activities in the primary market, maintain control over diverse operating units, and make strategic plans for all markets. Accordingly, costs rise when horizontal growth becomes excessive and organizational change becomes necessary.

3.4 Multidivisional Organizational Structure

Several economists (Keren and Levhari, 1979; Williamson, 1975) have suggested that the span of control limits the time of the top manager. Further, Radner (1978) showed that managers facing both short-run and long-run problems may focus

---

[2]Williamson (1975) asserts that managerial performance can be compared to a rate of return measure, a similar description

[3]Several economists (Lazear and Rosen, 1981; Silver and Auster, 1969; and Lucas, 1978) have argued that the executive manager is a more able decisionmaker than a subordinate manager.

attention on short-run plans. Hence, as the span of control imposes time demands that approach the constraint of managerial time, a top manager may focus on short-run planning rather than long-run planning, resulting in insufficient control over subsidiaries and inadequate strategic planning.

Eliminating the time constraint on top management requires either a change in business strategy or a restructuring of the organization. Change can come about by either reducing secondary market activities or hiring a monitor to oversee the primary line of business. If firm management decides to eliminate or limit secondary activities, then the firm takes on either the functional or F.S. forms. Alternatively, if management chooses to hire a monitor to oversee daily decision making in the primary market, then the firm adopts the multidivisional form.

For multiproduct firms, change occurs from diversification into new products or geographic markets. The multidivisional form accommodates the change by growing laterally in new divisions, thereby enabling a reduction in information travel time from top to bottom or vice versa. Moreover, because its sub-units are usually organized on a functional basis in which new functions are added laterally, it adapts to change in the product environment by growing laterally at the sub-unit level.

According to Williamson (1975), the multidivisional form is an organizational framework that enables a firm to more efficiently process information and control opportunism. As portrayed here, however, it enables better information processing than only multiproduct functionally organized firms and decentralized firms.

For diversified firms, the multidivisional form offers informational advantages in several ways. First, by decentralizing daily decisions into teams of managers whose performance can be measured relative to product performance, the multidivisional forms enables managers to obtain accurate information regarding subordinate work effort.[4] Second, by decomposing large centralized teams of managers into smaller sub-unit teams, managerial team sizes drop, allowing for a more precise measure of worker output (Alchian and Demsetz). Third, Shavell (1979) has shown that better measurement of work effort allows the firm to make better compensation packages, while Ouchi (1980) has argued that better measurement leads to better goal congruence between the worker and the firm. Fourth, Lazear and Rosen (1981) have shown that better information and competition for promotions allows the firm to sort higher quality managers from inferior ones. Fifth, the multidivisional form is structured such that the top manager can devote most of his effort to strategic planning while still monitoring subordinate performance. The cost, however, is that a less able monitor oversees the primary product lines.

---

[4]Lazear and Rosen (1981) have shown that under conditions of imprecise measurement of work effort, output-based pay scemes are superior to input-based scemes.

Kumar (1985) argues that the multidivisional form can lower the cost of merging with another unit. He suggests that its structure allows the firm to add a new merger partner as a sub-unit that can be made subject to divisional performance standards. Moreover, it can combine the managerial functions that are common to both itself and the new sub-unit and eliminate duplication of managerial work effort. The functionally organized firm, in contrast, must either decompose a merger partner into its functional parts and distribute them within its existing structure or add the new unit as a subsidiary. If it decomposes the existing firm into parts, then it must grow radially and be subject to delays in information travel time, while if it adds the unit as a subsidiary, then uncontrolled horizontal growth occurs, causing the firm to suffer from duplication of managerial effort and lose centralized direction.

Chapter 4

Firm Strategies and Histories to 1973

4.1 Overview

In the following chapter, historical data from the oil industry is applied to the model discussed in the previous chapter with particular emphasis placed on the necessary compatibility of firm strategy and organizational structure. For all of the firms in the sample, the period under study is a time of significant change. Firms that had previously chosen to be single-product and single-market firms began diversifying into other markets, while multiproduct and multimarket companies that had failed to consolidate their diverse holdings made organizational changes that enabled them to integrate their sub-units.

The incentives for each firm were similar and yet different. Each saw the multidivisional form as the best way to organize. Further, most firms saw benefits in international oil production and petrochemical production. Firms differed, however, as to whether diversification into secondary lines of businesses was necessary and, if so, the type of acquisition that was most profitable.

The common convergence to the use of the multidivisional form for different firm sizes and company business strategies prompts several questions. First, what were the organizational constraints that firm management faced? Second, how was one organizational structure able to accommodate diverse business strategies? In order to provide some insight to these questions, the following chapter presents brief company histories with an emphasis on the period surrounding the time of adoption of the multidivsional form.

4.2 Centralized Firms

All functionally organized firms in the sample adopted the multidivisonal form and changed business strategy from a single-product focus to geographical and product diversification between 1940 and 1970.

4.2.1 Ashland Oil Company. Prior to 1945, the Ashland Oil Company operated as a small oil refiner serving the Eastern Kentucky market. The company focused its marketing efforts on filling market niches not penetrated by the major oil companies while simultaneously exploiting the availability of low-cost river transportation for shipping products to their markets. Between 1945 and 1950, company management changed strategy, growing out of its traditional markets and into the Mid-Atlantic and Midwest, while increasing sales sevenfold to $145 million.

Ashland made its first move out of oil refining and marketing in 1958 when it began producing petrochemicals from refinery by-products. However, it was not until it purchased United Carbon in 1962, a carbon black and crude oil producer, that it moved more actively into diversified oil products and oil production. Diversification was carried one step further during 1964-67 when Ashland purchased several chemical firms, all of which were placed into a subsidiary under direct control of top management.

There was a change in organizational structure in 1967 when Ashland diversified its marketing activities more broadly. As a result, Ashland consolidated its chemical subsidiary into a division and devoted more resources to crude oil production and diversification into road construction and coal production.

The strategy behind Ashland's move into oil production and related oil products appeared to be twofold. First, it was dependent on outside sources of oil for most of its oil supplies and perceived a need to backward integrate in order to guarantee a source of supply. Second, its entry into the construction business was based on a strategy of forward integration into the end-user market for diversified oil products such as asphalt.

4.2.2 ARCO. Atlantic Refining (ARCO) was formed in 1911 after the dissolution of the Standard Oil Trust. As initially established, the company owned the east coast refining assets of the old Standard Oil Trust. The company's initial strategy, after its formation, was to integrate into transportation, service stations, and oil production. While it succeeded in building a transportation and marketing network, it was not able to increase its crude oil self-sufficiency rate beyond 50 percent until the mid-1960's.

After Robert Anderson became chairman in the 1960's, he changed both the strategy and structure of ARCO. To reach a new strategic goal of becoming a fully integrated nationwide firm, he merged the company with Richfield, a west coast refiner and marketer with large potential reserves in Alaska. He integrated Richfield into ARCO by making the marketing and refining units of Richfield a west coast division and consolidated production, exploration, and research and development into those functions at ARCO.

After the merger with Richfield, Anderson made plans to expand into the midwestern marketing and refining markets, develop petrochemicals, and diversify into mining (*Businessweek*, 1980a). His top priority was expansion into the midwestern market. After losing out to Sun in an acquisition battle over Sunray DX, Anderson acquired Sinclair Oil. The new acquisition gave ARCO the midwestern marketing presence that it sought, a petrochemical business, and additional acreage in Alaska.

4.2.3 CONOCO. Prior to 1911, Continental Oil (CONOCO) was affiliated with, but not directly controlled by the Standard Oil Trust. Its major role was that of a marketer of grease, kerosene, and oil products in the western states. As a result of its divestiture, CONOCO had neither refining nor production capabilities.

CONOCO's strategy immediately following its divestiture was to purchase a refiner in order to enter the gasoline and aviation fuel markets and secure a reliable source of products for its markets. In 1929, the oil company continued its backward integration with the purchase of Marland Oil Company, a firm with midwestern marketing outlets and significant oil reserves in the Southwest. In the 1930's, CONOCO integrated into research and development as a way to develop new fuels, petrochemicals, improved motor oils, and low cost refining technology. As a result of its development efforts, CONOCO became both a major aviation fuel producer and a significant petrochemical producer during World War II.

After World War II, in anticipation of a new growth strategy, CONOCO decentralized management control on a regional and product-line basis. Subsequently, CONOCO expanded into refined product markets on the East and West Coasts, began producing oil in the Middle East, and expanded its petrochemical businesses.

CONOCO began diversifying more broadly in the 1960's by entering bituminous coal and uranium markets. Both businesses required exploration skills that CONOCO possessed, and had similar markets and required a high degree of fixed investment but since operations were not redundant, direct cost advantages through consolidation did not occur and many skills were not transferable.

CONOCO's motivation to become a large diversified firm is unclear. Its initial diversification into highly related businesses apparently allowed CONOCO to use its existing talents in other product areas. For example, petrochemical diversification stemmed from research into petroleum products, while geographic diversification required an extension of existing talents into new locations. Later diversification efforts into coal and natural resources, however, had fewer similarities and thus had a lower potential for profitability.

4.2.4 Getty Petroleum. Prior to 1937. Getty Petroleum was an oil production firm operating in California. At this time, however, J.Paul Getty changed the business orientation of Getty Petroleum through the acquisition of controlling interests in Mission Oil, Tidewater Oil, and Skelly Oil, which were organized as majority-owned subisidiaries. Within this framework, Getty remained mainly a production firm that sold its output to its partially owned subsidiaries.

Growth also came in two nonpetroleum industries. First, in the Mission Oil transaction, Getty gained direct control over Spartan Aircraft and operated it as a subsidiary. Second, in 1938, Getty acquired the Hotel Pierse in New York, organizing it as a subsidiary. As a result, Getty Petroleum became a highly centralized holding company that owned majority interests in two other oil firms and had complete ownership over hotel and aircraft businesses.

After World War II, Getty Petroleum joined the search for oil in Saudi Arabia and Kuwait. It did not strike oil until 1953 and did not begin active production until 1954. Having oil supplies but needing a secure market for its crude, Getty Petroleum entered the Italian refining market in 1958 and supplied other oil to Tidewater for processing and sale in the eastern U.S. market.

Getty Petroleum's growth strategy also included entry into related businesses. During the 1950's Getty and Tidewater began actively exploring and developing uranium reserves. As a result, by the late 1950's, Getty was active in three different domestic businesses and foreign marketing and production but remained a relatively small centralized firm. Getty adapted its organization in such a way that J. Paul Getty remained the chief strategist, but all operating decisions were made by subordinates managing sub-units organized along product and geographic lines.

Throughout his tenure as chairman, J. Paul Getty maintained a strong aversion to debt and thus a fear of major acquisitions of any type. This fear of using leverage compelled Getty to maintain both Tidewater Oil and Mission Oil as subsidiaries in which Getty had controlling interest. In 1967, however, recognizing the benefits of a more unified structure, Getty took full control of Tidewater Oil but retained Mission Oil as a partially owned subsidiary.

4.2.5 Gulf Oil. After making huge oil strikes in Texas in the early 1900's, Gulf oil management acted to secure a market for its oil by forward integrating into refining. As a result, it became a refiner and marketer in the Southwest, South, and East Coast.

Gulf's first international activity began in the 1920's when it made oil discoveries ~~~~ela, Kuwait, Iraq, and Bahrain. Financial difficulties in the 1930's, however, required it to sell its interests in Iraq and Bahrain to Chevron and a share of its Venezuelan interests to Exxon. During this same period, Gulf slowly expanded into the midwest refined products market and adapted a functional organizational framework with control centralized at its Pittsburgh headquarters.

After World War II, Gulf's Texas oil fields began to lose their productive capacity, prompting Gulf to increase production at its Kuwait fields. After imposition of the Oil Import Quota by the U.S. Government, however, it was forced to increase domestic exploration and look for alternative outlets for its foreign crude. As a result, Gulf forward integrated into European marketing and refining, building small refineries and elaborate supply networks throughout Europe. At the same time, Gulf used technology that it had developed to enter petrochemical business.

In 1956 Gulf reorganized itself in order to adapt its structure to its diverse geographic and product markets. It chose to decentralize control of daily decision making to newly created geographic sub-units that were under direct centralized control.

According to Greene (1985), Gulf undertook a more far-reaching diversification drive in 1967 as a way to hedge against nationalization of its foreign production properties. As a result, Gulf diversified into nuclear reactors and land development, neither of which proved to be profitable.

4.2.6 Kerr-McGee. Before 1945, Kerr-McGee operated primarily as a contract driller. A change in strategy after World War II, however, enabled it to grow rapidly during the late 1940's and early 1950's. Management strategy was to expand as an integrated oil firm out of its southwestern base and to enter related mining businesses. Its first move out of the oil industry was its entry into the uranium market. This action was motivated by a desire to exploit its oil exploration skills in the prospecting for uranium, a belief that the uranium market would grow at the expense of oil, and the guaranteed profitability of the cost plus mark-up contracts offered by the Atomic Energy Commission (Ezell, 1979).

In order to adapt its structure for growth, Kerr-McGee streamlined its management in 1954 by creating a level of middle management that included managers for planning, refining and marketing, production, exploration, and uranium. Afterwards, Kerr-McGee accelerated its growth by expanding into all phases of both the oil and uranium businesses and entering the potash, phosphate, nitrogen, coal, and international oil markets.

4.2.7 Marathon. Before 1911, Marathon was an oil production subsidiary of the Standard Oil Trust. It remained only a crude producer until 1924 when it purchased Lincoln Oil and Refining Company, a small midwestern refiner. Its expansion, both in refining and production, led to a geographic market that by 1931 included Illinois, Indiana, Kentucky, Ohio, and parts of other midwestern states.

From the mid-1950's to the mid-1960's, Marathon expanded its domestic refining operations, entered into foreign markets, and adopted the multidivisional structure. Marathon's first major international oil strike occurred in 1958, but, because of the Oil Import Quota of 1959, it was unable to import foreign crude oil supplies. Consequently, Marathon expanded into European refining in 1961 as a way to secure a market for its Middle Eastern crude.

In the United States, Marathon merged with Aurora Oil in 1959, thereby doubling its capacity and obtaining petrochemical facilities. Later, it merged with Plymouth Oil in order to enter the southeastern market, acquire import and export facilities, and obtain a bigger oil import quota to help sell Libyan crude oil (Spence, 1962). Marathon integrated newly acquired firms and foreign marketing activities into its organizational structure, making these new units regional divisions. In the organizational arrangement that followed, Marathon centralized control over decentralized sub-units that were responsible for daily decision making.

After its transition to the multidivisional form, Marathon remained an oil and petrochemical firm and did not diversify into other businesses. Its main strategy was to concentrate on oil production while maintaining a profitable marketing operation. Its aversion to size at any cost is most evident in its European distribution network, where it departed first from the Italian market and, later, partially from other operations.

4.2.8 Occidental. Prior to the appointment of Armand Hammer in 1958 as chairman, Occidental existed as a small California oil producer. Under Hammer, business strategy changed to a drive for growth through diversification into natural resources, with a goal of eventually reaching the size of a major natural resource producer. By 1963, besides oil, Occidental had significant interests in sulfur, phosphorus, fertilizer, and ammonia and had reached a size of $77 million in sales. Throughout this period, the stock price rose from $10 in 1962 to $22 in 1965.

Occidental obtained a lucrative oil concession from Libya in 1966, giving it an opportunity for rapid growth. Shortly after striking oil, Occidental began producing about 500,000 barrels per day of Libyan crude, selling it on the spot market. Later, as a way to guarantee a market for its crude, Occidental bought the European marketing operations of Signal Oil.

According to *Forbes* (1968a), diversification efforts by Occidental into coal, inorganic chemicals, and land development were designed to act as a hedge against the possible loss of its Libyan production and as a means to exploit its tax credits. Moreover, as Occidental was diversifying its product line, it also sought to achieve a more geographical balance in production by expanding into Venezuela, Bolivia, Peru, Columbia, and the North Sea.

The organizational framework of Occidental established Hammer as the chief strategist, while sub-unit managers made most of the operating decisions. This framework enabled Occidental to readily integrate its acquisitions into its organizational structure by adding divisions in parallel to existing sub-units.

4.2.9 Sun Oil Company. Prior to 1900, Sun Oil acted as producer and supplier of gas oil products and other refined products to customers on the East Coast. It became an oil producer in Texas in the early 1900's as a way to supply its refining needs for production of gas oil, asphalt, and motor oil. Sun's strategy of vertical integration also led them into production of tankers during World War I, both for its own use and for sale to others.

Sun changed its strategy after World War I when it became a major refiner of gasoline. Despite its late entry, however, it became known as an innovative refiner, producing both no-lead high-octane gasoline and high grade aviation fuels. Moreover, rather than continue its focus on vertical integration, it chose to be dependent on outside suppliers because of the abundance of oil on the market. Hence, Sun became a crude-deficient oil refiner operating on the East Coast under the highly centralized control of the Pew family, an orientation it retained until the 1960's.

Sun's strategy changed in 1965, after the retirement of the eldest Pew. First, it expanded its geographic marketing area by merging with Sunray DX in 1967, giving it a 37 state marketing area. Second, it began actively exploring for oil, making self-sufficiency in crude oil its strategic goal. Third, it adopted a divisional structure that enabled decentralized control of a larger organization through three operating groups.

4.2.10 Tenneco. Tenneco was originally incorporated as the Tennessee Gas Transmission Company as a transporter of natural gas from Texas to the Appalachian area. In the 1950's, however, it began pursuing interests in oil and gas production and other buisnesses as a way to diversify into nonregulated businesses.

Throughout the 1960's, its strategy remained much the same. Tenneco management looked for poorly performing firms in any business, infused modern management techniques into their management systems, and then combined them with complementary firms in order to give them more breadth in the market. This strategy led to diversification into chemicals, packaging, land, cattle, farm and construction machinery, and auto parts. All new acquisitions were integrated into the firm either as a semi-autonomous division or as a division of one of the sub-units. In this framework, top management monitored divisional managers through assorted accounting and performance standards but removed itself from daily operating matters.

Although control was centralized, the eight divisions were only loosely connected and, hence, operated in a quasi-independent manner. While the independent nature of the markets that the divisions served resulted in offsetting a balanced cash flow for the entire company, investors did not place a premium on the company. In 1970, for example, the stock traded at nine times earnings, a value that, according to *Dun's* (1970), was too low for the types of businesses in which Tenneco operated.

4.2.11 *Texaco.* The Texas Company (Texaco) obtained huge deposits of readily extracted oil in Texas at the beginning of the 1900's. Its original strategy was to sell as much oil as possible and thus take advantage of its low cost source of oil. In order to achieve this goal, Texaco grew into a national distributor, while it vertically integrated into refining and marketing. Later, after expanding into foreign production and needing an outlet for its foreign oil, it formed Caltex with Chevron as a way to sell oil products in Europe. Finally, after realizing significant growth into several geographic markets, Texaco decentralized management in 1941 by extending daily decision making to geographic sub-unit managers.

After World War II, Texaco followed a strategy of developing its foreign properties and establishing its own European marketing network while remaining fully integrated. Unlike many of its competitors, Texaco did not attempt to diversify outside of the oil and petrochemical business. Rather, it focused on building small inexpensive refineries able to process only its own low-sulfur crude oil supplies. This was successful in an era of high transportation costs and ready access to low-sulfur crude oil, but by the late 1960's domestic reserves were becoming depleted and foreign sources were being nationalized. Moreover, the industry moved toward high-efficiency large volume refineries because of a decrease in the relative cost of transportation to refinery size. As a result, Texaco's profitability declined by the late 1960's.

4.2.12 *UNOCAL.* Prior to 1931, Union Oil (UNOCAL) existed as a California production firm. Afterward, because the company began feeling pressure from dropping crude prices, company management forward integrated into marketing and refining as a way to secure a market for its crude oil. This change in business strategy to an emphasis on refining changed the company's relative position in crude oil supplies from a seller on the crude market to a net buyer, as its self-sufficiency ratio dropped to 60 percent by the 1950's. The company accommodated the change in business strategy by replacing its entrepreneurial organizational structure with a functional arrangement.

In 1962, UNOCAL made a significant change in strategy toward one of growth. It first reorganized itself into a collection of profit centers with centralized control. Then, in an attempt to enter the East Coast market, it made an unsuccessful attempt to merge with Richfield Oil Company. Later, after a bidding war with other firms, UNOCAL merged with Pure Oil, thereby becoming a nationwide firm. In the reorganization that followed, Pure's refining and marketing units were made a midwestern division' while production, research and development, and exploration were combined with those same units at UNOCAL. Finally, because it lacked crude oil, UNOCAL shifted strategy toward crude self-sufficiency, which it attained in 1970.

4.3 Decentralized Firms

For decentralized firms, the multidivisional form provided a way to centralize control over diverse operations.

4.3.1 *Amoco.* Before 1911, Standard Oil of Indiana (Amoco) was the marketing and refining arm of the Standard Oil Trust in the Midwest. It later devoted its efforts toward improving the efficiency of its refineries and expanding in its Midwest marketing area. Amoco began to integrate backward into production and research and development during the 1920's. By 1937, the company had attained a 27 percent self-sufficiency ratio and had developed several new chemicals.

By 1951, Amoco was operating in the 41 states east of the Rockies through seven subsidiaries that included Standolind in the Midwest, Utah Refining in the Southwest, Pan American Southern Corporation in the South, and Pan American in the East. Subsidiaries acted independently except for centralized control over capital expenditures.

In 1955, Amoco acted to centralize control by first merging Pan American into Standolind and then later consolidating all of its subsidiaries, unifying marketing, refining, marketing and transportation activities. As a result, top management shifted its focus away from the Standolind subsidiary and towards firm policy decisions, planning, and firm-wide control.

After adopting the multidivisional form, Amoco's strategy became one of producing more domestic oil, expanding abroad, and increasing chemical sales. By 1971, Amoco had increased its self-sufficiency ratio to 75 percent from 47 percent in 1960, enabling it to free itself from outside purchases of oil in some of its major markets. During that same time frame, it changed from having no interests in overseas markets to having 25 percent of its assets abroad and increased its petrochemical sales by 600 percent.

4.3.2 *Chevron.* Before 1911, Standard Oil of California (Chevron) was a refining and marketing arm of the Standard Oil Trust. After divestiture in 1911, however, Chevron began to emphasize a search for crude oil in California and in foreign countries. Overseas, its first major source of crude oil came with the Bahrain concession purchased from Gulf. Later, an even more valuable concession was bought from King Saud of Saudi Arabia. At that point, Chevron had large foreign reserves but no refining operations. Accordingly, it formed Caltex with Texaco in 1935 to act as a European marketing outlet for its Middle Eastern oil output.

The mid-1950's was a period of rapid growth. In foreign markets, Chevron developed reserves in Bahrain, entered the Japanese and other Asian markets, and formed ARAMCO with Mobil, Exxon, and Texaco to gain access to Saudi Arabian oil. At the same time, Chevron began marketing east of the Rockies for the first time. Chevron was also active in chemical research and production, becoming the second largest producer of petrochemicals in the United States through its chemical division. In response to its growth into new markets, management at Chevron centralized control through a headquarters that was responsible for all strategic planning and control of division sub-units, which were based on geography in the oil business and product line in other businesses.

4.3.3 *Cities Service*. Prior to the mid-1930's, Cities Service was both an oil firm and an owner of public utilities. Legislation in the 1930's, however, compelled firms to either divest themselves of nonregulated businesses or sell their public utility holdings. Cities Service chose to sell its utilities, acquire interests in metals, and undertake oil refining and marketing, while organizing itself as a holding company.

A shift in strategy and structure occurred in 1962. Cities Service first consolidated its diverse holdings and then attempted to more rapidly diversify into products outside of the oil industry. Its new strategy led to growth into fertilizers, carbon black, plastic housewares, copper mining, and other miscellaneous activities. These new businesses, however, failed to be profitable and many were sold by the late 1960's.

4.3.4 *Exxon*. Exxon emerged from the 1911 break-up of the Standard Oil Trust primarily as a refiner and marketer of petroleum products on the East Coast and international markets. Because of its dependency on crude oil, Exxon immediately established two subsidiaries, Carter and Louisiana, as production companies. Through these subsidiaries, Exxon was able to increase its domestic crude production to refined product ratio from 14 per cent in 1918 to 32 per cent in 1927. During this time, Exxon also grew geographically into the midwestern and southern states.

In order to manage its growth efficiently, Exxon adopted a centralized F.S. organizational structure in 1927 with the creation of five semi-autonomous multifunctional operating units and one subsidiary managed by the central office. Within this framework, top management had ample time for both strategic planning and monitoring all of its domestic subsidiaries. Its foreign operations, however, remained organized along functional lines, spanned multiple geographic jurisdictions, and were not directly monitored by central headquarters.

Exxon's long-term strategy was to grow as an integrated oil firm, both domestically and abroad. This focus led Exxon to search extensively for oil, resulting in it gaining access to low cost oil in Saudi Arabia, Venezuela, Iraq, and other foreign countries and eventual self-sufficiency in domestic oil. Moreover, integration into research and development enabled it to develop new refined-oil products, petrochemicals, and petroleum cracking technology.

In the early 1960's, Exxon again reorganized operations and changed strategy. The oil business had matured by this time, and Exxon had gained many new competitors both abroad and in its domestic markets. Its organizational structure in the 1950's, however, did not allow it to quickly respond to market changes. Moreover, Exxon still had not consolidated and reorganized its foreign interests. As a result, Exxon first consolidated its domestic subsidiaries into a single unit and then established regional organizations headed by general managers in overseas markets that were directly monitored by central headquarters.

Under its new organizational structure, Exxon made its first venture outside of oil and petrochemicals. Company president Michael Hader indicated that Exxon diversified because it became concerned about the risks of oil company nationalizations and the possibility of oil resources eventually becoming exhausted (*Businessweek*, 1965). Accordingly, Exxon undertook geographic diversification of its oil production properties and began exploring for uranium and coal. Its emphasis, however, was to remain in the oil business.

4.3.5 *Mobil Corporation*. Mobil Corporation is a descendent of two firms that merged in 1931, Standard Oil of New York (SOCONY) and Vacuum Oil Company. Before the merger, SOCONY was primarily a marketer of refined oil products in 29 eastern and southwestern states; the Vacuum Oil Company was a marketer and producer of high quality motor oil. The combined company was strong in refined product sales, both domestically and abroad, but weak in crude oil supplies. Nonetheless, it had acquired large reserves of oil in the Middle East as a consequence of a partnership with Exxon.

According to Greene (1985), the two companies were never completely integrated. In a reorganization in the late 1950's, Mobil established divisions that were based on geographic and product markets, created a planning and coordinating department, established standardized accounting methods, and fully integrated all subsidiaries into the firm.

After its reorganization, Mobil created a European marketing network that could sell its low cost Middle Eastern oil. It diversified into petrochemicals in order to more efficiently use refinery by-products and it vertically integrated into plastics markets as a way to use its petrochemical products.

4.3.6 *Phillips Petroleum*. Phillips Petroleum was a major producer of crude oil and natural gas in the 1920's, supplying both Amoco and Exxon. When its two main customers backward integrated into production, it was forced to forward integrate into refining and marketing, a focus that it retained until the mid-1960's. By that time,

Phillips' concentration on domestic marketing and refining led to growth into all of the states and, with expertise it developed in petrochemicals, expansion into petrochemical production that generated 20 percent of its revenue.

Management strategy in the mid-1960's was best enunciated by the president of Phillips, Bill Keeler, who said the following (*Businessweek*, 1968b):

> Phillips has often found minerals it simply didn't know what to do with....Our problem is how to turn these materials into dough, regardless of what they are....It's putting a whole new life into our minerals group. They're running around like a bunch of turpentined cats.

This strategy led to unorganized growth with market activities that ranged from plastic products to metals. Moreover, management failed to make similar investments in oil production and as a result did not produce a sufficient supply of crude oil to meet its refining needs.

Up until the early 1970's, the company was organized such that top managers decided policy by committee while subordinates at subsidiaries made their own strategic and operating decisions. At this time, Phillips' established a planning department, shifted to a profit-center management structure, appointed outside directors, and adopted centralized control over subisidiaries.

Within this new framework, Phillips established a new strategy of consolidation and increased crude production. As a result, Phillips sold its plastics end-user businesses and its metals businesses, exited the west coast gasoline retailing market, and reduced employment by 10 percent. It then redirected investment towards oil exploration, becoming self-sufficient by 1975.

4.3.7 *Shell Oil.* Shell Oil Company was established by Royal Dutch Shell in the early 1900's as a subsidiary operating in the western United States market. Its most significant growth period came during the 1920's when it expanded into all of the Pacific Coast and adjoining states and expanded into both the Midwest and East Coast. Shell accommodated its growth by establishing independent subsidiaries for all three of its marketing regions.

In the 1930's, Shell became an active researcher in refining technology and petrochemicals and also acted to consolidate operations within each subsidiary. After World War II, because it had excess wartime capacity and had developed considerable expertise in chemical research and development, Shell was well positioned to grow in the petrochemical business.

A centralization program was undertaken in 1949 as a way to consolidate three affiliated subsidiaries into one centralized company. The consolidation process went through three phases until a final arrangement was eventually reached. In the initial phase, top management merged all domestic oil units into one company and established a headquarters in New York. In the second phase, management organized chemical, oil production, pipeline, and refining sub-units into semi-autonomous subsidiaries. In the last phase, managers achieved complete integration when it merged all of its subsidiaries into the parent company. Within this framework, Shell primarily concentrated on chemicals and oil.

4.3.8 *Standard of Ohio*. Prior to its 1911 divestiture from the Standard Oil Trust, Standard of Ohio acted as a refined products marketer in Ohio. Its strategy for its first 45 years after its divestiture was to remain primarily a refiner and marketer in Ohio. The company made a marked departure from its past in 1959 by developing a diversification program and making a complete overhaul of its corporate organization. In the reorganization, Standard first cut its payroll by 10 percent and then, because it wanted to be able to diversify into other products while being able to accurately evaluate the performance of new operating units, replaced its functional with subsidiary organizational framework with a product-oriented profit-center arrangement.

The growth that Standard of Ohio sought came through diversification into plastics, oil shale, petrochemicals and motels, and growth into regional gasoline markets near its Ohio base. Further, in an effort to become self-reliant in oil production, Standard traded 53 percent of its stock to British Petroleum in 1969 in exchange for their Prudhoe Bay holdings in Alaska. For Standard of Ohio this marked a significant break from its traditional refining and marketing base and into oil production.

4.4 Discussion and Conclusion

In the sample of oil firms discussed in this chapter, there were two basic types of organizational arrangements prior to the multidivisional era. One type, the highly centralized structure, could not readily accommodate growth. Hence, diversification built upon this framework was costly. The other type, a decentralized type, was a result of disorganized growth in which planning was largely decentralized, controls were not maintained, and redundancies existed across operating units.

One group of centrally organized firms (companies with functional or corrupt multidivisional organizational forms) adopted the multidivisional form as a way to manage growth into diverse geographic and product markets and to facilitate mergers. As shown in table 4.1, five of these firms entered international oil markets and seven diversified into either chemicals or other business lines or both around their transition period. If this type of growth was handled along functional lines, it would have been necessary for top managers to monitor subordinate managers that were responsible for markets outside their area of expertise. Moreover, if one manager were responsible for two products then imprecise measurement of work effort may have resulted if superior performance in one business offset poor effort in another.

A second group of four functionally organized firms adopted the multidivisional structure as a way to better consolidate merger partners in the oil industry. Typically, these firms retained their partners' refining and marketing operations as a division and consolidated other functions into their own operations.

A final group of eight firms, all decentralized firms (F.S. or holding company structures), adopted the multidivisional form as a way to consolidate growth from a prior period. Most of these firms had been active in all regions of the country, foreign markets, and petrochemicals prior to adopting the multidivisional form. They originally created independent subsidiaries to handle most decision making in secondary product or geographic markets, but since many of these markets were similar to the primary business, some duplication of effort existed. Moreover, most planning efforts were devoted to one market while others were left to either languish or grow randomly. Hence, for these firms, the multidivisional form offered a way to consolidate their holdings and to grow in a more centrally directed way.

The strategies followed by most oil firms in the sample were remarkably similar. Many of the firms had made efforts to vertically integrate, entered international markets, and were producing petrochemicals by 1975, all of which may have appeared profitable at the time of entry. International oil, for example, provided an opportunity to find low-cost oil, vertical integration enabled firms to avoid an opportunistic supplier, and petrochemical production allowed firms to use refinery by-products as inputs for chemicals.

Growth outside of local markets appeared to have been in a sequential pattern. Initial growth was by entry into adjacent regions and international markets. Later, firms added petrochemical complexes to refineries in order to realize higher mark-ups on their oil and then grew into other energy businesses, some activity that was a secondary outcropping from the oil business or into unrelated businesses.

TABLE 4.1–NEW BUSINESS ACTIVITIES UNDERTAKEN BY COMPANIES
AT THE TIME OF TRANSITION[1]

Firm[*]	Merger	Centralized Control	Foreign Oil	Chemicals	Other
Ashland Oil (F)				x	
ARCO (F)	x				
CONOCO (F)				x	
Getty Oil (M')			x		
Gulf Oil (F)			x	x	
Kerr-McGee (F)					x
Marathon Oil (F)	x				
Occidental (F)			x	x	x
Sun Oil Co. (F)			x	x	
Tenneco (F)					x
Texaco (F)			x		
UNOCAL (F)	x				
Amoco (FS)		x	x	x	
Chevron (FS)		x	x		
Cities Service (H)		x		x	
Exxon (FS)		x			
Mobil Corp. (FS)		x		x	
Phillips (H)		x	x		
Shell Oil (FS)		x			
Standard Ohio (FS)		x	x	x	

F=functional form; FS= functional with subsidiary form; H=holding company form; and M'=corrupted multidivisional form.

[1] Top half of table contains centralized firms; bottom half contains decentralized firms.

[*] Organizational form prior to transition to multidivisional form in parenthesis.

Chapter 5

Empirical Models

5.1 Overview

Two types of empirical tests were run to test alternative beliefs of the motivation behind adopting the multidivisional form. The first uses event study methodology to evaluate the wealth effects of the multidivisional form. The second tests for changes in profitability, managerial costs, and sales.

The diversification theory, as Chandler interprets it, makes four suggestions, of which the first three differ from Williamson (1975) and Armour and Teece (1978). First, multidivisional firms do not realize a higher level of profitability than single-product functionally organized firms. Second, functionally organized firms will adopt a multidivisional form only when they intend to diversify or have already done so. Third, large F.S. organized firms adopt the multidivisional form as a way to reduce redundant managerial capacities across subsidiaries and to give the top manager more time for managing secondary activities. Fourth, holding company firms are inherently decentralized; hence, adopting the multidivisional form provides them with a centralized means of control. Table 5.1 gives a summary of predicted outcomes of Armour and Teece (firm size) and Chandler (diversification) theories of the multidivisonal form. Table 5.2 summarizes the sign predictions for each empirical test.

5.2 Wealth Effects

The discussion of alternative organizational forms indicates that there were costs associated with decentralized control. Insufficient control is inherent to holding companies, but for F.S. forms, centralized control depends on the number of subsidiaries and the complexity of the primary business line. One proxy for the number of subsidiaries may be the overall size of the firm, while another might be the degree of diversification. Hence, if the number of subsidiaries is proportional to both size and amount of diversified sales, then firms with these attributes should have more subsidiaries and, as a result, be more decentralized.

If the F.S. and holding company forms are decentralized and thus inefficiently organized, then the decision to change to a centralized organizational structure should signify an increase in profitability. Moreover, if increasing profitability results in greater dividends, then investors should be made wealthier. Therefore, investors should bid up the price of the stock of a decentralized firm on the day of the announced change. Investors in single product functionally organized firms, however, may not bid up stock prices because functionally organized firms are already efficiently organized.

Armour and Teece (1978) believe that the multidivisional form is a superior organizational form to the functional form. Hence, investors of functionally organized firms would be expected to bid up stock prices on the day of the announcement.

5.2.1 *Event Studies.* Investor response to firm reorganization plans was measured with two types of event study tests. First, daily or monthly data were used to measure stock price changes on the day or month of the announcement of the change in organizational form. The use of daily data is commonly accepted as the preferred choice because it limits the potential for conflicting announcements, as is possible with monthly tests. The data, however, are available from publicly available sources only in the post-1962 period. Accordingly, the approach followed here was to use daily data when it was available and monthly data, which dates to 1929, otherwise. Equally weighted stock data was used for both tests.

Second, Cumulative Average Residual (CAR) methodology was used to test for investor wealth gains under conditions of event date uncertainty. Two situations caused uncertainty: first, multiple announcements were made by several firms; second, it was possible that firms could make unannounced operating changes that were discerned by some investors prior to the announcement. An advantage of CAR methodology is that it does not require a precise announcement date, but a drawback is that standard significance results lose their applicability because of the strict conditions required for statistical analyses. Hence, CAR methodology is used in this book as qualitative evidence.

5.2.2 *Event Date Selection.* The date selected was the first firm-wide centralization, decentralization, restructuring, or reorganization announcement appearing in either the *Wall Street Journal, Businessweek,* or *New York Times* and consistent with

---

[1]Tests for the effects of consolidations at both Cities Service and Ashland Oil were made with both firms having insignificantly negative results. These were dropped because consolidations may signal to investors that bad times are ahead.

the transition period (the period of adjustment for the firm as it changes organizational form) given by Armour and Teece (table 5.3).

Fourteen firms and 17 announcements are identified in table 5.4. Three announcements, those of Cities Service and the initial announcements by Amoco and Ashland Oil were dropped because they were termed consolidations which, while being essential for multidivisionalization, may provide a misleading signal to investors.[1] The last two announcements by Exxon, marking the adoption and then return to a functional organizational form by its domestic operating unit, were also dropped because of an earlier announcement. Kerr-McGee was not listed on the NYSE and could not be tested. Hence, the announcements of management restructurings by Amoco and Ashland Oil and the announcement of reorganizations plans by Exxon were used while both Cities Service and Kerr-McGee were dropped from the sample.

Other complications included: multiple announcements by six firms, management changes accompanying the announcements by six firms, and a move of company offices by one firm. Neither management changes nor a move of company offices are inconsistent with multidivisionalization but multiple announcements can cause event date selection ambiguity. If the firm made multiple announcements, the one used was regarded as the first descriptive evidence of the intention to change organizational form.

5.2.3 *Monthly and Daily Tests.* Standard event study methodology, which tests the significance of the regression residual in Equation 5.1, was used for both the daily and monthly tests where:

$$(5.1) \quad e_{j,t} = R_{j,t} - a_t - B_{j,t} R_{m,t}$$

where

$e_{j,t}$ =abnormal return or residual to stock "j" in month or day "t",
$R_{j,t}$ =realized rate of return to stock "j" in day or month "t",
a_t =intercept term from Ordinary Least Squares,
$B_{j,t}$ =beta of security "j", and
$R_{m,t}$ =return on market portfolio in month "t".

Market model parameters for the daily tests were estimated for the 180 trading days beginning 190 days prior to the announcement. Abnormal stock performance was then estimated over the three-day period centered on the announcement day. Market model parameters for the monthly tests were estimated for the 90 months beginning 91 months prior to the announcement month. Abnormal stock returns were measured at the end of the announcement month.

5.2.4 *Cumulative Average Residual.* Under conditions of event date uncertainty, CAR methodology can be used to provide qualitative evidence of wealth gains. The construction of the test follows Schipper and Thompson (1983). It aligns 22 firms in event time with their event date, either the first announced change (table 5.4) or January of the first year of the transition period (table 5.3).[2] Several steps were followed. First, market model parameters were estimated for each firm by means of the window technique used by Schipper and Thompson. They estimated the market beta from the 48 months surrounding the announcement date (equation 5.2). Second, average residuals for the portfolio of firms were obtained by summing all abnormal residuals and dividing by the number of firms (equation 5.3). Third, cumulative average residuals were then summed across time for a period starting 50 months prior to the announcement and ending 90 months afterward (equation 5.4).

If wealth effects do exist for firms changing organizational form, then the cumulative average residual should begin to rise at or prior to the event year and continue rising until leveling off some time beyond. The difference between its final peak and its minimum prior to the event year should reflect the wealth effects.

$$(5.2) \quad R_{j,t} = a_{j,t} + B_{j,t} R_{m,t} + e_{j,t} \qquad t = t-24, t+24$$

$$(5.3) \quad AR_t = \sum_{j=1}^{J} e_{j,t}/J_t$$

$$(5.4) \quad CAR_t = \sum_{t=1}^{T} AR_t$$

where

$R_{j,t}$, $a_{j,t}$, $B_{j,t}$, $R_{m,t}$, $e_{j,t}$ are the same as above,

AR_t = average residual for the sample during month "t",

J_t = the number of firms in the sample during month "t", and

CAR_t = cumulative average residual starting at time "t" and extending over a range from 70 months prior to 100 months after the announcement.

5.3 Models of Operating Performance

Event studies suggest that potential gains (losses) for investors exist but do not reflect an actual change in performance. Accordingly, other models of performance differences are used for further testing.

5.3.1 *Cross-Sectional Tests of Long-Run Profitability.* Many economists believe that at any point in time, firms within any given industry should realize a similar rate of long-run profitability. Accordingly, deviations from average long-run profitability for the industry may reflect economic rents associated with differences in either technology, resource costs, research and development, or organizational efficiency.

---

[2]Schipper and Thompson used the last announcement, but the first announcement was used here because it was the most relevant information available to investors at that time.

Salinger (1984) used Tobin's Q, the ratio of the market value of the firm to the replacement value of its physical assets, as a measure of long run profitability. Short-run measures of profitability taken from accounting data can show wide variations if costs are poorly estimated or accounting rules change. For example, if costs are ten times profits and costs are mismeasured by one percent, then profits are mismeasured by 10 percent but if capital stock is mismeasured by 1 percent then Tobin's Q is mismeasured by only 1 percent.

Pooling time-series and cross-sectional models can be used to test for profitability differences, but two problems arise. First, a control variable for variations in Tobin's Q with overall market expectations must mimic investor behavior towards all oil firms and vary with the market. One such portfolio can be constructed for firms that do not change organizational form; however, it is identical for all firms in the sample and introduces statistical complications. Second, the period under study is a time of significant institutional, technological, and market changes. Hence, models evaluating profitability across time may reflect institutional differences rather than organizational ones. As a result, annual cross-sectional tests were made for each year from 1950-72.

In the cross-sectional model, several variables can be used to capture differences in profitability among firms. Teece (1976) argues that oil firms vertically integrate as a way to avoid the risk associated with opportunistic suppliers and marketers. Accordingly, long-run profitability for vertically integrated oil firms may be higher than for non-integrated firms. Hence, a test for organizational efficiency must control for the degree of vertical integration.

Another source of economic rents for oil firms can stem from the gains associated with low-cost sources of oil. Burrows and Domencich (1970) assert that 1957 Middle Eastern oil imports were $1.50 per barrel cheaper than domestic oil. Further, Sampson (1975) indicates that Middle Eastern countries granted oil firms long-term contracts on the most productive tracts of land. Hence, firms that held these contracts may have earned economic rents.

A third effect on oil company profitability may have resulted from the Oil Import Quota of 1959, which changed the relative profitability of foreign-produced oil. Burrows and Domencich (1970) assert that, because of lower production costs overseas, imported oil during the 1960's could have realized a $1.25 profit over domestic oil, if it were permitted to enter the country. Because of the quota, however, foreign oil had to be sold abroad. Hence, because domestic and overseas markets were independent, it may have been efficient to vertically integrate in both foreign and domestic markets.

Other sources of economic rents may stem from exploiting research and development in new markets. Teece (1976), for example, indicates that Standard of

Ohio earned $600 million from one patented petrochemical process. Further, Owen (1985) claims that contracts granted by the Atomic Energy Commission during the 1950's enabled oil firms to realize a profit of $3 to $7 on one pound of uranium. Moreover, Ezell (1979) asserts that oil firm exploration skills can be efficiently exploited in the search for uranium. Hence, economic rents may have arisen from entry into these markets, but then would have dissipated as other firms entered the market.

A fifth effect on oil company profitability may have been induced by the change from one organizational form to another. Williamson indicates that a firm changing organizational arrangements must undertake a transition stage consisting of a series of organizational changes. Hence, an accurate comparison of an efficient structure to an inefficient one should control for the transition period. Armour and Teece (1978) identified this period for 19 oil firms (table 5.3). The sample under study here includes the Armour and Teece specification for 17 firms and extended transition periods for Shell and Cities Service, which had begun consolidating before the time given by Armour and Teece. Further, because Armour and Teece gave no transition period for three firms in their sample, Texaco, CONOCO, and Kerr-McGee, the relevant period was obtained from the *New York Times*, *Wall Street Journal*, *Businessweek*, and company histories.

The final effect on oil company profitability stems from its organizational form. Accordingly, dummy variables were created for three types of organizational forms: functional (includes all centralized forms), decentralized (includes both holding company and F.S. forms), and multidivisional. The functional form dummy contains all companies that were organized as functional firms and the one firm that was organized as a centralized corrupted multidivisional. The decentralized form includes all functional subsidiary firms and holding company firms. Finally, the multidivisional form encompasses all multidivisional firms.

Measures that were used to capture the effects discussed above are as follows. First, Teece (1976) believes that the ratio of oil production to refinery needs reflects the degree of vertical integration for a given oil firm. Therefore, a ratio was used for both domestic and foreign-produced oil as a measure of the degree of vertical integration. Second, the importance of foreign oil production to firm performance was captured by the ratio of foreign oil production to total oil production. Third, dummy variables were used to capture the effect of organizational form. Fourth, since economic rents from research and development and exploration cannot be controlled, the types of markets and potential gains from research in which firms operate must be considered in any discussion of results.

Equation 5.5 is a model that specifies Tobin's Q as a function of domestic and foreign vertical integration, the amount of foreign-produced oil, and organizational form dummy variables. If, as Armour and Teece (1978) believe, the multidivisional form is more profitable than the functional form, then b_7 should be significantly positive. Moreover, if the transition period is also more efficient than the functional period, b_5 should also be significantly positive. If, on the other hand, the multidivisional form and single-product functional firms are equally efficient, then b_7 should be insignificant from zero. Moreover, if decentralized firms are less efficient than both multidivisional and single-product functional firms, then b_6 should be significantly less than zero. Further, if the transitional period is a costly period for functional firms, but perhaps more efficient for decentralized firms, b_5 should be insignificant from zero.

$$(5.5) \quad TOBINSQ_{it} = b_1 FFORM_{it} + b_2 FORVI_{it} + b_3 REFVI_{it} + b_4 FORIN_{it} + b_5 TRANS_{it}$$
$$+ b_6 DECENTRL_{it} + b_7 MLDV_{it} + e_{it}$$

where
$TOBINSQ_{it}$ = Tobin's Q for firm "i" at time "t",
$FFORM_{it}$ = intercept term and set equal to one throughout the period and represents all functional firms and all corrupted multidivisional forms,
$FORVI_{it}$ = measure of vertical integration in the U.S. market and defined as barrels of foreign oil produced divided by barrels of foreign oil refined,
$REFVI_{it}$ = measure of vertical integration in the U.S. market and defined as domestic oil production divided by domestic refining,
$FORIN_{it}$ = control variable for the degree of foreign oil products and defined as foreign production divided by total production,
$TRANS_{it}$ = dummy variable defined as one during the transitional period and zero otherwise,
$DECENTRL_{it}$ = dummy variable defined as one for decentralized periods and zero otherwise,
$MLDV_{it}$ = dummy variable defined as one for multidivisional periods and zero otherwise, and
e_{it} = independent and normally distributed dummy variable with a mean of zero.

5.3.2 *Long-Run Profitability in Time Series Test for Individual Firms.* Since cross-sectional tests cannot be adapted to control for firm performance over time, it is not known whether the more profitable firms have high earnings because of their organizational form or if they adopt an organizational form because they are more profitable. One way to control for firm history is to compare the performance of the company over time to a portfolio of other firms that do not change organizational form. If profitability differences can be attributed to organizational form then the multidivisional period should be significantly more profitable.

Lustgarten and Thomadakis (1987) indicate that Tobin's Q varies over time. As a result, time-series tests must control for fluctuations in the stock market. Accordingly, a portfolio of other firms in the industry that did not change organizational form was created as a "market portfolio" against which the firm under study could be

compared. Dummy variables can then be used to capture the effect of multidivisionalization.

Equation 5.6 is the specification of the model that will be used to evaluate the
performance of different firms during their multidivisional period relative to their
original organizational form period. If only functionally organized firms are more
efficiently organized as multidivisional firms, then b_{10}, b_{11}, or both should be positive
only for those firms while, if only decentralized firms are more efficiently organized
as multidivisional firms, then b_{10} or b_{11} should be positive only for those firms in the
following equation:

$$(5.6) \quad TOBINSQ_t = b_8 AVGQ_t + b_9 FFORM_i + b_{10} TRANS_i + b_{11} MLDV_i + e_t$$

where
$TOBINSQ_t$ = Tobin's Q for firm "i" at time "t" and includes the twenty periods
surrounding the transition period,
$AVGQ_t$ = average Tobin's Q of a portfolio of firms that do not change organizational formand includes the twenty periods surrounding the transition period,
$FFORM_t$ = intercept term defined as one during the entire test period,
$TRANS_t$ = dummy variable defined as one during the transitional period and
zero
otherwise,
$MLDV_t$ = dummy variable defined as one during the multidivisional period and
zero otherwise, and
e_i = independent and normally distributed dummy variable with a mean of zero.

   5.3.3 *t-Tests of Changes in Long-Run Profitability for Alternative Portfolios* .
If a firm loses its efficiency just prior to its transition period, the effect of multidivisionalization may not be reflected in the regression. If this is the case, a t-test that
compares the profitability of a firm just prior to its transition period to its profitability
at some point afterward should reveal a significant change in performance.

Equation 5.7 is a specification of relative Tobin's Q, which is defined here as
Tobin's Q at some point after its transition period relative to Tobin's Q of a portfolio
of other firms that did not change their organizational form. If the multidivisional
form generates more wealth than the functional form, then a change in the mean of
relative Tobin's Q should be significantly positive for functionally organized firms
that change their structure. Alternatively, if only large decentralized firms are inefficiently organized, then it should be significantly positive only for those firms.

$$(5.7) \quad RLTVTBQ_i = TBQ_i / PRT_j$$

where

RLTVTBQ$_i$ = difference in relative Tobin's Q, which is defined as Tobins's Q for firm "i" relative to Tobin's Q for a portfolio of control firms "J", for selected years before and after the transition to the multidivisional period,
TBQ$_i$ = Tobin's Q for firm "i", and
PRT$_J$ = average Tobin's Q for a portfolio of control firms "J" containing all firms that do not change organizational form during the test period.

5.3.4 *Means Test of Long-Run Profitability.*

An alternative test may be to compare the means of portfolios of alternatively organized firms over time. If the multidivisional form is more efficient than the functional form, the means of the multidivisional firms should be significantly greater than both the mean of the control group and the mean of the functionally organized firms that eventually change organizational form. Alternatively, if decentralized firms can be efficiently organized only with a multidivisional structure, then their mean should be significantly less than the mean of the multidivisional firms.

5.3.5 *Portfolio Comparison of Average Long-Run Profitability.*

Another way to compare profitability differences for different organizational structures is to make comparisons of portfolios of specific organizational forms. Since Tobin's Q should be equivalent across firms, a regression of one portfolio type on another should result in a coefficient of approximately one. Any differences from one would suggest a difference in profitability among portfolio types.

Equation 5.8 compares average Tobin's Q for a portfolio of multidivisional firms to average Tobin's Q for a portfolio of functionally organized firms in a control group that do not adopt the multidivisional form. If, as Armour and Teece (1978) indicated, the multidivisional form is more profitable than the control group, then b_{13} should be greater than one when b_{12} is approximately zero. Alternatively, if single-product functionally organized firms and multidivisional firms are equally efficient then b_{13} should be equal to approximately one.

$$(5.8) \quad AVGQMDF_t = b_{12} + b_{13}AVGQCNTL_t + e_t$$

where
$AVGQMDF_t$ = average Tobin's Q for a portfolio of all multidivisional firms,
$AVGQCNTL_t$ = average Tobin's Q for a portfolio of all functionally organized firms that do not change organizational form, and
e_t = independent and normally distributed dummy variable with a mean of zero.

Equation 5.9 is a specification similar to equation 5.8 in that it regresses average Tobin's Q for multidivisional firms on average Tobin's Q for functionally organized firms that later convert to the multidivisional structure. If large functionally organized firms are inefficiently organized, then b_{15} should be greater than one when the intercept is approximately zero.

$$(5.9) \quad AVGQMDF_t = b_{14}+b_{15}AVGQFNL_i+e_t$$

where
$AVGQMDF_t$ = average Tobin's Q for a portfolio of multidivisional firms,
$AVGQFNL_t$ = average Tobin's Q for functionally organized firms, and
e_t = independent and normally distributed dummy variable with a mean of zero.

Equation 5.10 gives a specification similar to equation 5.10, but regresses a portfolio of multidivisional firms on decentralized firms. If decentralized firms are inefficiently organized then b_{17} is predicted to be greater than one given small values of b_{16}.

$$(5.10) \quad AVGQMDV_t = b_{16}+b_{17}AVGQDCNTRL_i+e_t$$

$AVGQMDV_t$ = average Tobin's Q for a portfolio of multidivisional firms,
$AVGQDCNTRL_t$ = average Tobin's Q for a portfolio of decentralized firms with less than 10 percent of sales in diversified market
e_t = independent and normally distributed dummy variable with a mean of zero.

5.3.6 *Operating Profits.* Another way to evaluate the performance of the multidivisional form is to compare accounting measures of operating performance. Separate standard Ordinary Least Squares (OLS) regressions were run for each firm to evaluate the initial affect of the adoption of the multidivisional form on firm profitability. Since time series data were used, the first differenced log form was employed.

Operating profitability is evaluated in equation 5.11. If, as Armour and Teece (1978) assert, the multidivisional form is more efficient than a functional organizational structure for large firms, then the predicted sign on b_{22} is positive only for functionally organized firms. Moreover, if, as Armour and Teece also suggest, the transitional period is a profitable period, then b_{21} should be positive for that time. Finally, since Armour and Teece (1978) also believe that size determines organizational form, a ranking by firm size should reveal stronger results for larger functionally organized firms than for smaller firms. Alternatively, if only large decentralized firms are inefficiently organized then the predicted sign on b_{22} is positive only for those firms. Further, if some firms are able to improve efficiency during the transition period, they may realize a gain in profitability in that period. Accordingly, for any given large decentralized firm b_{21}, b_{22}, or both are predicted to be positive.

$$(5.11) \quad DERNING_t = b_{18}DSALES_i+b_{19}DOILP_i+b_{20}FFORM_i+b_{21}TRANS_i+b_{22}MLDV_i+e_t$$

where

DERNING$_t$ = first differenced log of operating profits for the 20 periods surrounding the transition period where profits = sales - cost of goods sold - general and administrative costs - depreciation - depletion,

DSALES$_t$ = first difference of logs of sales in time "t" for the 20 years surrounding the transition period,

DOILP$_t$ = first difference of log of oil price index at time "t",

FFORM$_t$ = intercept term that represents the original organizational form period and defined as one during the entire test period,

TRANS$_t$ = dummy variable defined as one during the transitional period and zero otherwise,

MLDV$_t$ = dummy variable defined as one during the multidivisional period and zero otherwise, and

e$_t$ = independent and normally distributed dummy variable with a mean of zero.

5.3.7 *Managerial Efficiency.* Kumar (1985) argues that the multidivisional form provides an efficient mechanism for absorbing merger partners. Functionally organized firms, however, must merge by either establishing a subsidiary, assigning functional components to its own functional structure, or decentralizing all business interests and adopting a holding company form. If it creates a subsidiary or adopts the holding company form there may be duplication of centralized management, thereby increasing administration costs. Hence, large decentralized firms and rapidly diversifying functionally organized firms should realize a reduction in managerial costs per unit of input after adopting the multidivisional structure.

Functionally organized firms that are not rapidly diversifying but are converting to the multidivisional form may not realize a reduction in managerial costs. If the firm was centrally controlled and begins staffing divisions, no duplication of work effort may exist. Hence, managerial costs per unit of output may or may not drop.

Equation 5.12 provides a specification from which the degree of consolidation after a change in organizational form can be ascertained. It gives a regression of general and administrative costs on barrels of refined oil. If a higher level of managerial efficiency occurs starting during either the multidivisional period or the transitional period, then either b_{25} or b_{26} should be negative for both large decentralized firms and functionally organized firms that make diversifying acquisitions prior to their multidivisional period. Functionally organized firms that did not recently diversify, however, would not be expected to realize a gain in managerial efficiency because they must create new divisional staffs.

$$(5.12)\quad DLGAEXP_t = b_{23}DOIL_t + b_{24}FFORM_t + b_{25}TRANS_t + b_{26}MLDV_t + e_t$$

where

DLGAEXP$_t$ = first differenced log of general and administrative costs for the 20 periods surrounding the transition period,

DOIL$_t$ = first differenced log of refined barrels of oil produced daily for the 20 periods surrounding the transition period,
FFORM$_t$, TRANS$_t$, MLDV$_t$ = same as above.
e$_t$ = independent and normally distributed dummy variable with a mean of zero.

5.3.8 *Sales Changes in Diversified Markets.* If one motivation to adopt the multidivisional form is to devote more time to secondary markets, then sales may rise in those markets. According to Wilson (1978), better information processing, when combined with constant returns to scale in technology, can allow the firm to grow. However, Williamson (1975) argues that better information may motivate the firm to reduce sales. Thus, for single-product firms and those with a small portion of sales in diversified markets, sales in diversified markets should rise; for highly diversified firms, however, sales in diversified markets may or may not rise.

Attention to secondary markets requires that less time be devoted to primary markets. Fewer resources devoted to those markets may result in a decline in sales. Hence, an increase in sales in secondary markets may be offset by a decline in sales in the primary market. Therefore, aggregate sales may or may not rise.

Kumar (1985) argues that the multidivisional form lowers the cost of mergers because merger partners can be more rapidly consolidated by adding them as a division. Hence, some firms may adopt the multidivisional form as a low-cost way to merge with other firms.

If the transition period to a multidivisional form is a costly period, a rational manager would change firm organizational structure only if the firm was adapting to a recent change in its focus on secondary markets, was currently changing its marketing focus, or was adjusting its organizational framework to accommodate a future change. Therefore, a four-period model becomes necessary: the first period concerns the original organizational form and business strategy; a second period, before the transition, captures those firms changing business strategy prior to the beginning of a change of structure; a third period, during the transition, covers those firms simultaneously changing structure and business strategy; and a final period allows for firms that may change business strategy after their structural change. The four periods are specified as follows: the transition and multidivisional periods are as given by Armour and Teece (1978), the period before the transition was assumed to be the three years before the transitional period, and the original organizational form period was assumed to reach 10 years beyond the period before the transition (table 5.3).

Equation 5.13 gives a specification of sales growth into diversified markets. If managers of single product firms, those with less than 10 percent of sales in diversified sales, change organizational structure in response or anticipation of future growth into new product markets, positive signs are expected on b$_{29}$, b$_{30}$, b$_{31}$, or any combination.

Large diversified firms, those firms with diversified sales in excess of 10 percent of sales and \$100 million before the transition, may choose to diversify more under the multidivisional structure or decide to consolidate their operations in diverse markets. Therefore, previously diversified firms may have positive or negative coefficients.

$$(5.13) \quad DVRSFD_t = b_{27}DDVRSFD_t + b_{28}FFORM_t + b_{29}PRIOR_t + b_{30}TRANS_t + b_{31}MLDV_t + e_t$$

where

$DVRSFD_t$ = first difference of diversified sales during year "t" during the twenty years surrounding the transition period,

$DDVRSFD_t$ = lagged first difference of diversified sales for the twenty years surrounding the transition period,

$PRIOR_t$ = dummy variable defined as one during the three year period prior to the transitional period and zero otherwise,

$FFORM_t$, $TRANS_t$, $MLDV_t$ = same as equation 5.13, and

e_t = independent and normally distributed dummy variable with a mean of zero.

Predicted results are summarized in table 5.2. Armour and Teece predictions are classified as size motives while diversification motives are identified as diversification.

TABLE 5.1–EMPIRICAL PREDICTIONS FOR COMPETING HYPOTHESES

Test*	Test Type	Size	Diversification
Wealth Effects			
Functional	Event study	Wealth gain	No change
F.S.	Event study	No prediction	Wealth gain
Holding company	Event study	Wealth gain	Wealth gain
Long-Run Profitability			
Functional	Cross-section	Division higher	No difference
F.S.	Cross-section	No prediction	Division high
Holding company	Cross-section	Division higher	Division high
Long-Run Profitability			
Functional	Time series	Division higher	No change
F.S.	Time series	No prediction	Division higher
Holding company	Time series	Division higher	Division higher
Operating Profitability			
Functional	Time series	Division higher	No change
F.S.	Time series	No prediction	Division higher
Holding company	Time series	Division higher	Division higher
Managerial Efficiency			
Nondiversified functional	Time series	No prediction	No change
Diversified functional	Time series	No prediction	Division higher
F.S.	Time series	No prediction	Division higher
Holding company	Time series	No prediction	Division higher
Change Diversified Sales			
Nondiversified firms	Time series	No prediction	Division higher
Diversified firms	Time series	No prediction	No prediction

* In later sections, holding company and F.S. forms are combined into a decentralized category.

TABLE 5.2–SUMMARY OF SIGN PREDICTIONS FOR EACH EMPIRICAL TEST

Test and Firm Type	Firm Size	Diversification
Wealth Effects		
Functional	$e_{jt} > 0$	$e_{jt} = 0$
Decentralized	No prediction	$e_{jt} > 0$
Long-Run Profits (cross-section)		
Multidivisional	$b_7 > 0$	$b_7 = 0$
Transition	$b_5 > 0$	$b_5 = 0$
Decentralized	No prediction	$b_6 < 0$
Long-Run Profits (time series)		
Functional	b_{10} or $b_{11} > 0$	b_{10} or $b_{11} = 0$
Decentralized	No prediction	b_{10} or $b_{11} > 0$
Long-Run Profits (t-test) (time series)		
Functional	Positive	Insignificant
Decentralized	No prediction	Positive
Long-Run Profits (portfolio comparison)		
Divisional vs. control	$b_{13} > 1$	$b_{13} = 1$
Divisional vs. functional	$b_{15} = 1$	$b_{15} = 1$
Divisional vs. decentralized	$b_{17} = 1$	$b_{17} > 1$
Operating Profits (time series)		
Functional	b_{21} or $b_{22} > 0$	b_{21} and $b_{22} = 0$
Decentralized	No prediction	b_{21} or $b_{22} > 0$
Managerial Efficiency (time series)		
Single-product functional	No prediction	$b_{26} = 0$
Diversified functional	No prediction	$b_{26} < 0$
Decentralized	No prediction	$b_{26} < 0$
Diversified Sales (time series)		
Single-product functional	No prediction	b_{29}, b_{30}, or $b_{31} > 0$
Multiproduct functional	No prediction	No prediction
Single-product decentralized	No prediction	b_{29}, b_{30}, or $b_{31} > 0$
Diversified decentralized	No prediction	No prediction

TABLE 5.3–SAMPLE OF FIRMS AND THEIR ORGANIZATIONAL FORMS

Firm	Initial Form	Initial Strategy	Initial Period	t-Form Period	Final Form	Final Strategy	Final Period
Ashland Oil	F-form	reg	1930-66	1967-69	M-form	rel....un	1970-90
ARCO	F-form	reg	1930-63	1964-65	M-form	rel....un	1966-90
Cities Service	H-form	divest utilities	1944-62	1963-66[2]	M-form	rel....un	1967-82
CONOCO[1]	F-form	reg	1930-48	1949-51	M-form	rel....un	1952-81
Exxon	FS-form	rel	1930-59	1960-65	M-form	rel....un	1966-90
Getty Oil	C-form	rel	1930-58	N.A.	M-form	rel....un	1959-84
Gulf Oil	F-form	reg	1930-55	1956-57	M-form	rel	1958-84
Kerr-McGee	F-form	reg	1944-53	1954-55	M-form	rel	1955-83
Marathon Oil	F-form	reg	1930-59	1960-62	M-form	rel	1963-82
Mobil Oil	FS-form	merge	1930-58	1959	M-form	rel....un	1960-90
Occidental	F-form	reg	1962-67	1968-71	M-form	rel....un	1972-90
Phillips	H-form	rel	1930-72	1973-74	M-form	rel	1975-90
Shell Oil	FS-form	rel	1930-48	1949-59[2]	M-form	rel	1960-90
Chevron	FS-form	rel	1930-53	1954	M-form	rel....un	1955-90
Amoco	FS-form	rel	1930-55	1956-60	M-form	rel....un	1961-90
Standard of Ohio	FS-form	rel	1930-59	1960-61	M-form	rel....un	1962-90
Sun Oil Co.	F-form	reg	1930-67	1968-70	M-form	rel....un	1971-85
Tenneco	F-form	reg	1950-58	1959-61	M-form	un	1962-90
Texaco[1]	F-form	merge	1930-39	1940-42	M-form	rel	1943-90
UNOCAL	F-form	reg	1930-61	1962-63[3]	M-form	rel....un	1964-90

F-form (functional form): Functionally organized firm. FS-form (functional-with-subsidiaries form): Functionally organized firm in the primary market with subsidiaries in the secondary markets. C-form (corrupted multidivisional form): A divisionalized structure in which the top manager makes all decisions. T-form (transitional form): The transition period from one organizational form to the multidivisional form. M-form (multidivisional form): Divisionalized structure in which the top manager makes all strategic plans and division managers make operating decisions. H-form (holding company form): Divisionalized structure in which there is little centralized control. Reg: regional growth; merge: growth through merger; rel: growth into international oil, petrochemicals, coal, uranium; unrelated: growth into activities other than oil or related industries.

[1] Transitional and functional period based on company histories and Bhargava (1972).
[2] Differs from Armour and Teece because of a firm consolidation in 1963.
[3] Announcement data indicates that divisionalization began in 1962.

Source: Armour and Teece (1978), company histories, *Wall Street Journal Index*, and *Businessweek* articles.

TABLE 5.4–EVENT DATE ANNOUNCEMENTS

Firm	Announcement Date	Announcement Description
Ashland Oil	11-17-69	Reorganized and named a new CEO
	11-16-67	Consolidated chemical units
Cities Service	01-03-63	Consolidated
CONOCO	06-29-50	Decentralizated; moved offices to Texas
Exxon	12-14-60	Reorganized Humble by region
	11-12-65	Restructured Humble back to functional
Gulf Oil	03-11-57	Decentralized and changed management
Kerr-McGee	0-01-6-54	Created a new layer of management
Marathon Oil	04-04-61	Divisionalized domestic operating units
Mobil Oil	01-13-59	Divisionalized and shifted management
Phillips	01-15-74	Reorganized firm and promoted eight executives
AMOCO	09-29-60	Revealed reorganizational plans
	02-01-56	Consolidated chemical units
Sun Oil Co.	09-17-69	Reorganized firm and named a CEO
Tenneco	12-12-60	Partially reorganized
Texaco	10-31-40	Reorganized
UNOCAL	04-30-62	Modified company; shifted management

Note: ARCO, Getty Oil, Occidental, Chevron, and Standard of Ohio made no announcement of management changes.

Chapter 6

Data

Stock value, accounting, and organizational form data were required for this study. The stock value data was obtained from CRSP tapes published by the University of Chicago, while accounting data were obtained from *Moody's Industrial Manual*, and organizational form data were acquired from Armour and Teece (1978), company histories, and Bhargava (1972).

The Armour and Teece organizational form data (table 5.3) contains information on the organizational form of 27 firms listed in the 1975 Fortune 500 with more 50 percent of their revenues from the oil business. For 19 of these firms, the data includes the time periods for their original, transitional, and multidivisional organizational form. For three other firms, CONOCO, Kerr-McGee and Texaco, only the final multidivisional period was provided. As a result, the timing of their transition period was obtained from company histories and company announcements. Another firm, Clark Oil, begins its transition in 1975 and shortly afterward was sold. Accordingly, this firm was not included in the time series tests. The final four firms in the sample did not change to the multidivisional form. In summary, the sample contains 22 firms with original, transitional, and multidivisional periods and five firms that did not change their organizational arrangements.

Tobin's Q was computed by using an approach similar to Lindenberg and Ross (1981). The market value of the firm was assumed to include common stock, preferred stock, current liabilities, and debt, while assets of the firm were assumed to include inventory, property, plant, and equipment, and all other assets.

The components of the market value of the firm were computed as follows. The market value of common stock was assumed to be the average of the beginning and ending market values, while both preferred stock and current liabilities were valued at book value. Finally, long-term debt was computed by discounting it to the present. The discount rate was assumed to be identical with Moody's composite average of industrial bonds.

The asset accounts of property, plant, and equipment, and inventories also required adjustments. Oil industry accounting requires that property be reported as the value of estimated proven reserves, leased producing properties, unproven reserves, land for downstream operations, and the costs of drilling productive oil wells. Historical values were adjusted to current values by assuming that all equipment was new in 1930 or, if the firm was not listed in Moody's until a later date, that it was new at its initial listing. It was further assumed that property, plant and equipment depreciates at a five percent yearly rate. Accordingly, current year value was made equal to 95 percent of the previous year's value of property, plant, and equipment value and was adjusted for the GNP deflator plus the change in property, plant, and equipment during that year.[1] This approach is similar to that used by Smirlock, Gilligan, and Marshall (1984).

Inventory adjustments depended on the accounting method used. When first-in and first-out (FIFO) inventory accounting was used, the historical value was assumed to be identical to the book value. But when other accounting methods were used, the replacement value was assumed to be equal to the book value times the current price of oil divided by the oil price at the beginning of the year.[2]

Some theorists (Lustgarten and Thomadakis, 1987; Salinger, 1984) believe that research and development, oil exploration costs, and advertising should be considered the costs of acquiring intangible assets and therefore included in the denominator of Tobin's Q. For this study, however, data on such expenditures were not available. Hence, Tobin's Q should be interpreted as the market value of a firm's tangible assets, an approach used by Smirlock, Gilligan, and Marshall (1984).[3]

The long-run profitability in time series test required a variable consisting of average Tobin's Q for a portfolio of firms to serve as a control for stock market fluctuations. Firms included in the portfolio were those firms that did not change organizational form during the period studied. The composition of portfolios is identified in the relevant table where it is used. Summary statistics for all portfolios

---

[1]The GNP (Gross National Product) deflator reflects price changes in the agrregate economy.

[2]Oil prices were used because most inventory was assumed to be an oil product.

[3]Since all firms are in the same industry, it is reasonable to assume that firms devote similar amounts to research and development and advertising. Hence, any overstatement of firm Q is common to all firms and does not affect the results.

are presented in table 6.1, while summary statistics for individual firms are presented in Table 6.2.

Statistics reflecting the size of each firm in the 20 years surrounding the time of transition to the multidivisional form is indicated in table 6.3 and is used for ranking firms by size. As can be noted, there is significant variation in sales volume between the smallest and largest firm.

Not all data was available for managerial efficiency and sales in diversified markets tests. Three firms, Chevron, Standard Oil of Ohio, and Kerr-McGee, did not report general, sales, and administrative expenses and were dropped from the sample testing managerial costs. Further, since Occidental organized as a company in 1959 and its initial results may be misleading, all of its data up to 1962 were dropped. Three firms Kerr-McGee, Texaco, and Gulf, were dropped from the diversified sales study because of insufficient data, while estimates for international oil sales at Mobil, Marathon, Exxon, Chevron, and Amoco were determined from sales of barrels of refined oil given by *Moody's Industrial Manual* and refined oil price data published by the Bureau of Labor Statistics.

The definition of sales in diversified markets was taken to be sales outside the primary market of the firm. As a result, oil sales in international markets and sales outside the oil market were considered diversified sales. For two coastal firms, ARCO and UNOCAL, diversified sales were assumed to be their growth by merger into the East and West Coast oil markets.[4] The data for sales into diversified markets were obtained from SEC filings, annual reports, and *Moody's Industrial Manual*.

---

[4]Mergers and sales increases were assumed to take place at the initiation of merger talks.

TABLE 6.1–DISTRIBUTION OF TOBIN'S Q FOR ALTERNATIVE PORTFOLIOS

Portfolio	Mean	Standard Deviation	Minimum	Maximum
A*	1.273	0.343	0.690	2.020
B	1.015	0.239	0.715	1.820
C	1.130	0.179	0.746	1.530
D	1.061	0.124	0.758	1.290
E	1.084	0.223	0.634	1.430
F	0.785	0.233	0.432	1.109
Control*	1.369	0.657	0.530	3.540
Decentralized	1.000	0.229	0.605	1.611
Functional that changed to multidivisional	1.210	0.424	0.666	3.161
Multidivisional	1.170	0.368	0.465	2.352

Portfolio A = American Petrofina, Commonwealth, Superior, Texaco; Portfolio B = Superior, Texaco, Phillips, CONOCO; Portfolio C = Sun, Sunray, Superior, Phillips; Portfolio D = Ashland, Phillips, Sun, Sunray; Portfolio E = Ashland, ARCO, Superior, Marathon, UNOCAL, Sun, Sunray; Portfolio F = Ashland, ARCO, Getty, Marathon, UNOCAL; Control = all firms that do not change organizational form; Decentralized = all holding company and F.S. forms; Functional = all functionally organized firms that change to the multidivisional form; and Multidivisional = all multidivisional firms.

* Portfolio used for the cross-sectional test and the individual firm tests.

TABLE 6.2–DISTRIBUTION OF TOBIN'S Q

Firm	Mean	Standard Deviation	Minimum	Maximum
American Petrofina	0.864	0.214	0.530	1.222
Ashland	1.111	0.240	0.762	1.607
ARCO	0.885	0.258	0.667	1.883
Belco	1.410	0.409	0.684	2.115
Cities Service	0.823	0.138	0.647	1.225
Clark Oil	1.327	0.699	0.671	3.162
Commonwealth Oil	1.306	0.387	0.723	1.921
CONOCO	1.369	0.464	0.709	2.340
Exxon	1.217	0.237	0.739	1.661
Getty Oil	1.252	0.590	0.572	2.648
Gulf Oil	1.119	0.200	0.761	1.442
Kerr-McGee	1.540	0.282	1.063	2.285
Marathon Oil	1.272	0.166	0.899	1.552
Mobil Corp.	0.918	0.142	0.605	1.130
Murphy	0.898	0.199	0.623	1.367
Occidental	1.312	0.558	0.492	2.354
Phillips	1.142	0.146	0.865	1.511
Richfield	1.158	0.126	0.953	1.428
Shell	1.390	0.311	0.846	2.136
Chevron	1.248	0.237	0.807	1.710
AMOCO	0.884	0.146	0.666	1.149
Standard Oil of Ohio	1.011	0.236	0.738	1.507
Sun Oil Company	1.155	0.282	0.699	1.650
Sunray	0.970	0.108	0.722	1.150
Superior	2.042	0.674	1.146	3.540
Tenneco	0.879	0.132	0.629	1.161
Texaco	1.438	0.435	0.751	2.352
UNOCAL	0.850	0.166	0.464	1.255
United Refining	1.124	0.258	0.866	1.610

TABLE 6.3–DISTRIBUTION OF SALES AND BARRELS OF REFINED OIL
FOR INDIVIDUAL FIRMS[1]

Firm	Mean Sales (1967 prices)	Mean Refined Oil (barrels)	Standard Deviation of Sales	Standard Deviation of Oil
Belco	85.77	21.90	81.46	14.47
Kerr-McGee	171.00	80.00	138.00	66.30
Getty Oil	259.10	60.14	475.38	108.33
Clark	316.23	89.24	161.31	22.48
Murphy	344.19	96.95	200.61	34.59
CONOCO	492.05	118.38	229.74	55.47
Marathon Oil	501.52	104.76	217.50	58.58
Standard Oil of Ohio	606.58	174.05	255.89	84.00
UNOCAL	837.55	247.95	489.25	119.33
Tenneco	901.47	56.47	757.93	23.64
Texaco	1,094.78	314.48	446.70	96.97
Cities Service	1,274.73	321.00	151.23	31.78
Ashland Oil	1,355.47	255.10	927.08	86.50
Chevron	1,408.69	403.33	592.39	117.98
ARCO	1,532.55	433.19	1,259.30	295.52
Shell Oil	1,986.00	549.00	1,035.00	234.00
Sun Oil Company	2,064.35	446.19	1,348.75	157.57
Occidental	2,229.25	232.45	1,647.46	137.74
AMOCO	2,423.91	734.24	399.88	165.57
Gulf Oil	2,663.76	762.62	877.96	293.39
Phillips	3,190.83	588.33	1,474.08	91.29
Mobil Corp.	3,669.84	1,033.19	1,491.48	440.03
Exxon	12,684.41	3,781.48	5,321.72	1,177.29

[1] Ranked by mean sales volume.

Chapter 7

Results

Table 7.1 summarizes the empirical findings. Tables 7.2 and 7.3 give event study results for functionally organized firms and decentralized firms. None of the functionally organized firms realized an abnormal gain on the day of the announcement. Moreover, three of the nine firms realized negative returns.

Two decentralized firms, Amoco at the five percent level of significance and Phillips at the ten percent level, realized abnormal gains in profitability on the day or month of the announcement. Moreover, all decentralized firms realized a positive gain on the day or the month of the announcement. The negative residual on the day prior to the announcement for Phillips and the month after for Exxon confounds their results. The negative residual at Phillips, however, may have been caused by a dividend announcement on that day. Nonetheless, the weak results of both samples suggest that either investors do not significantly benefit from a change in organizational form or that considerable leakage of information existed.

Figure 7.1 shows an erratic average stock price throughout the test period for functionally organized firms.[1] A stock value appreciation of 15 percent begins at a point 16 months before the event month and peaks eight months prior to the event year, but drops 10 percent over the following 17 months. It then rises once again by 11 percent over the next 12 months before declining 19 percent in the next 35 months.

Figure 7.2 indicates that for decentralized firms there was no peak prior to its multidivisional period. The pattern that follows, however, is strikingly similar to that of the functionally organized firms. After the hundredth month there is a stock value

---

[1]See Table A.1.1 for numerical values for the period studied.

increase of 10 percent, then a 19 percent drop, and finally an erratic pattern until the end of the test period.

Many plausible explanations for the cyclical pattern can be offered. For example, the initial increase in stock price for functional firms may have been from investor response to the decision of the firm to enter businesses in which oil firms may have had a comparative advantage. The later decline could then be the realization that diversification can be costly. Further, the increase afterwards by both groups may have resulted from the reorganization. Finally, the subsequent cyclical pattern could have been caused by the reaction of investors to economic events in the oil industry or investors reacting to firm diversification efforts (table 7.4). Regardless of the explanation, however, the overall pattern suggests that, if there were gains to organizational change, they were small.

Table 7.5 contains results for cross sectional tests of long-run profitability for every year between 1950 and 1972. Models were run in log form in order to correct for heteroskedasticity and the Least Absolute Residual method (LAR) was used to correct for outliers (Maddala, 1977).[2] Table 7.6 contains results for the uncorrected models, while table 7.7 has results after removal of the outliers. A Durbin-Watson test rejected autoregressive behavior for seventeen of the periods, while six fell in the inconclusive region.

The results of table 7.5 indicate that decentralized firms significantly underperformed centralized firms (functional and corrupted multidivisional firms) in 14 periods at the 90 percent level and 19 at the 80 percent level.[3] At no time did decentralized firms outperform multidivisional firms. Multidivisional firms underperformed centralized firms in five periods and outperformed them in three.

One plausible explanation for these results may stem from the marketing activities in which each firm was engaged. Table 7.8 indicates that, in the earliest period, decentralized firms and multidivisional firms were active in similar markets, while centralized firms were mainly domestic oil firms. In later periods, centralized firms remained single-market firms, multidivisional firms remained multimarket firms, and decentralized firms were an intermediate case. Multidivisional firms may have outperformed centralized firms in the earliest period because of economic rents from their early entry into international markets, petrochemicals, and uranium and outperformed decentralized firms because of a superior organizational efficiency. During the 1960's, however, entry occurred into all these markets, thereby depressing profitability and resulting in an elimination of economic rents. The weak performance

---

[2]A Goldfield-Quandt test failed to reject the null hypothesis that no heteroskedasticity was present.

[3]Getty, which employed a highly centralized divisional structure, the corrupted multidivisional structure, was the only centralized firm in this group that was not functionally organized.

in the late 1960's and early 1970's period may have arisen from events in international markets and from government actions embodied in the Williams Act (1968), Williams Amendment (1970), and the 1969 Tax Reform Act. According to Schipper and Thompson (1983), all of these laws made diversification more costly.[4]

The time series regression results of long run profitability are reported in table 7.9. After adjustment for auto-correlation, a Durbin-Watson test rejects auto-correlation error for all firms but Kerr-McGee, which falls in the inconclusive region. Table 7.9 indicates that three centralized firms realized a gain in profitability, and six had a significant reduction in profitability during either the transitional or multidivisional period. Seven decentralized firms realized a gain in profitability during either the multidivisional or transitional period, and none recorded a decline in profits.

T-test results in time-series are reported in table 7.10. This table indicates that there was an insignificant drop of Tobin's Q across a 20-year period for all firms.[5] The results, however, differed according to the degree of decentralization, size, and time of organizational change. After segmenting the data according to the degree of decentralized control, Table 7.10 indicates that there was an insignificant drop in Tobin's Q for all centrally organized firms and an insignificant gain for decentralized firms. Moreover, after eliminating all firms that changed after 1965, Table 7.10 shows that there was an insignificant increase in Tobin's Q for all firms. Finally, by comparing the eight largest decentralized firms with the eight largest centralized firms, one can note that there was an insignificant increase in Tobin's Q for all large firms.

Results for means tests of alternative portfolios are reported in table 7.11. The mean of the control portfolio is shown to be significantly greater than that of the multidivisional form during 1956-72, while the mean of the centralized firms that change their structure is insignificantly different from that of the multidivisional firms. Finally, the mean of the portfolio of multidivisional firms is significantly greater than that of the decentralized firms during 1951-72.

Results for long-run profitability tests comparing portfolio Tobin's Q are reported in table 7.12. The coefficients on both the control portfolio and the functional form portfolio are less than one while the coefficient on the portfolio of decentralized firms is greater than one, indicating that decentralized firms were less profitable than multidivisional firms and that centralized firms were more profitable than multidivisional firms.

---

[4]These events were initiated by Libya, which raised the price of oil that it sold to Occidental and later nationalized a large amount of their assets. Libyan action was soon followed by OPEC, of which Libya was not a member.

[5]Since the OPEC cartel distorted the valuation of company assets in the 1970's, the final year considered for all firms is assumed to be 1972 unless their transition was 1973, in which case 1979 was assumed to be the final year for the test period.

Operating profitability results are reported in table 7.13. Centrally organized firms are ranked according to size while decentralized firms are arranged by size and degree of diversification. Five of the 13 centralized firms have negative results, of which one is significant at the 90 percent or higher level. Eight firms report positive coefficients, two of which were significant at the 90 percent or higher level. Seven firms have insignificantly negative coefficients during the transitional period, while four have insignificantly positive coefficients.[6] For decentralized firms, only one firm reports a significantly positive result for either the multidivisional or transitional period. All firms, however, have positive coefficients during their multidivisional period.

Although tests of the two samples have mainly insignificant results, there are distinct differences. First, results for the decentralized firms are insignificantly positive during both periods studied for seven of the nine firms, while this was true for only four of the twelve centrally organized firms. Moreover, if significance levels of 20 percent or less are considered, the three largest decentralized firms are significant in one of the two relevant periods and five of the seven largest are significant in one of the two periods. For centralized firms, however, two of the four largest firms reported significantly positive results, while one has a significantly negative result. Further, only three of the nine largest firms have significantly positive results, while two are significantly negative.

Results for managerial efficiency tests are reported in table 7.14. Three decentralized firms have statistically significant improvements in managerial efficiency during the multidivisional period, while the other four have insignificant improvements.[7] Centrally organized firms have three significant improvements in managerial efficiency during one of the two periods and two significant increases in managerial costs.

Results for diversification efforts are reported in table 7.15. Eleven of the 13 firms with less than 10 percent of sales in diversified markets prior to their transition have significantly positive results. Marathon, one of the remaining two firms, did report a significant gain in sales. This increase in sales reflects its growth through mergers with Aurora Oil and Plymouth Oil companies and its geographic diversification into international oil. Shell, the remaining firm, used its transition to consolidate its domestic operations and manage growth into petrochemicals. Furthermore, two firms that could not be tested, Gulf and Kerr-McGee, also expanded diversified sales (table 7.15). Gulf expanded into chemicals and international oil, while Kerr-McGee expanded from its regional base in oil and began to diversify into uranium production,

---

[6]Tests regressing earnings on organizational form showed no significant difference from the regression on sales.

[7]Regressions of sales and general and administrative expenses on sales yield a similar result.

eventually becoming the largest uranium producer in the United States. Hence, by combining these results, fifteen firms may have been motivated to adopt the multidivisional form as a way to adapt their organizational structure to a changing business strategy.

TABLE 7.1–SUMMARY OF EMPIRICAL FINDINGS

Test	Firm Size[1]	Diversification[2]
Wealth Effects (event study)	Fails to reject null hypothesis	Fails to reject null hypothesis
Long-Run Profit (cross section)	Fails to reject null hypothesis with or without outlier control	Rejects null hypothesis in 19 of 23 periods at 20 percent level
Long-Run Profit (time series)	Fails to reject null hypothesis	Rejects null hypothesis for seven of nine firms at the 5 percent level
Long-Run Profit (t-test)	Fails to reject null hypothesis	Fails to reject null hypothesis
Operating Profit (time series)	Fails to reject null hypothesis	Rejects null hypothesis in five of seven cases at the 20 percent level
Managerial Efficiency (time series)	No prediction	Rejects null hypothesis for largest firms
Diversified Sales (time series)	No prediction	Rejects null hypothesis for firms with less than 10 percent of sales in diversified markets

[1] The null hypothesis states that the multidivisional form allows better performance than the functional form.

[2] The null hypotheses states the multidivisional form allows higher profitability than decentralized forms, decentralized forms realize a gain in managerial efficiency, and single product firms adopt the multidivisional form as they are diversifying.

TABLE 7.2–MONTHLY AND DAILY RESIDUALS FOR CENTRALIZED FIRMS

Firm	Three-Day Window[1]	Event Month	Residual	Studentized Residual
Ashland Oil[2]	11-16-69	-	-0.0000	-0.0001
	11-17-69	-	0.0168	0.7640
	11-18-69	-	-0.0027	-0.1220
Sun Oil Co.[2]	09-16-69	-	-0.0029	-0.2410
	09-17-69	-	0.0175	1.4300
	09-18-69	-	0.0057	0.4770
CONOCO	-	06-30-50	0.0848	1.6620
Gulf Oil	-	03-31-57	-0.0087	-0.1718
Marathon Oil	-	04-30-61	-0.6200	-0.6200
Tenneco (33 obs.)	-	12-31-60	0.0410	1.1110
Texaco	-	10-31-41	0.1021	1.5800
UNOCAL	-	04-30-62	-0.0152	-0.3110

[1] The middle date is the announcement date. [2] Daily data used for Ashland and Sun.
* 10% level of significance. ** 5% level of significance. *** 1% level of significance.

TABLE 7.3–MONTHLY AND DAILY RESIDUALS FOR DECENTRALIZED FIRMS

Firm	Three-Day Window[1]	Event Month	Residual	Studentized Residual
Phillips[2]	01-14-74	-	-0.0641	-4.0030***
	01-15-74	-	0.0306	1.9100*
	01-16-74	-	0.0019	0.0740
Exxon	-	07-31-60	0.0343	1.0000
		08-31-60	-0.0141	-0.0250
Mobil Corp.	-	01-31-59	0.0271	0.0645
Amoco	-	09-30-60	0.0840	2.2700**

[1] The middle date is the announcement date; [2] Daily data used for Phillips.
* 10% level of significance. ** 5% level of significance. *** 1% level of significance.

TABLE 7.4–DIVERSIFIED SALES AS A PERCENT OF TOTAL SALES[1]

Firm	TRANS	Year Relative to Transition Year					
		-10	-5	-1	+1	+5	+10
Amoco	1956-60	8 (ch)	6 (ch)	6 (ch)	9 (ch)	16 (ch,io)	31 (ch,io)
ARCO	1964-65	0	0	0	40 (merger)	40 (merger)	70 (merger)
Ashland Oil	1967-69	0	0	46 (ch,cm)	44 (ch,c,io,cm)	27 (ch,c,io,cm)	41 (ch,c,cm,sh)
Belco	1969[2]	0	0	20 (c)	34 (c)	30 (c)	6 (c)
Chevron	1954	26 (ch)	27 (ch,io)	34 (ch,io)	35 (ch,io)	36 (ch,io)	38 (ch,io)
Cities Service	1963-66	1 (m)	1 (m)	19 (m,ch)	23 (m,ch)	27 (m,ch,pl,io)	16 (m,ch,io)
CONOCO	1949-51	N.R.	5 (ch)	3 (ch)	4 (ch)	12 (ch,io)	16 (ch,io)
Exxon	1960-65	41 (ch,io)	40 (ch,io)	43 (ch,io)	52 (ch,io)	50 (ch,io,c)	40 (ch,io,c,u)
Getty Oil	1959[2]	0	29 (io,u)	44 (io,u)	64 (io,u)	87 (io,u)	60 (io,u,ch)
Gulf Oil	1956-57	0	N.R.	8 (io,ch)	13 (io,ch)	25 (io,ch)	36 (io,ch,c)
Kerr-McGee	1954	N.R.	N.R.	N.R.	N.R.	N.R.	N.R.
Marathon	1960-62	0	0	0	0 (u)	7 (u,m,io)	14 (u,m,io)
Mobil Corp.	1959	12 (io)	27 (io)	22 (io)	26 (io,ch)	32 (io,ch)	42 (io,ch)
Murphy	1972[2]	1 (a)	1 (a,c)	6 (a,c)	6 (a,c)	7 (a,c)	6 (a,c)
Occidental	1968-71	0	87 (m,f)	44 (m,f,r)	67 (m,f,r,c,io,ch)	69 (m,f,r,c,io,ch)	44 (m,f,r,c,io,ch)

Phillips 1973-74	30 (ch,m)	30 (ch,m)	31 (ch,m)	26 (ch,m,io)	42 (ch,io,u,c)	35 (ch,io,u,cl))
Shell Oil 1949-60	8 (ch)	4 (ch)	4 (ch)	13 (ch)	11 (ch)	12 (ch)
St. of Ohio 1960-62	0	0	3 (ch)	1 (ch)	21 (ch,pl,c)	16 (ch,pl,c)
Sun Oil Co. 1968-70	N.R.	6 (sh)	7 (sh)	13 (sh,ch,io)	31 (sh,ch,io)	49
Tennec 1959-61	0	0	0 (sh)	18 (ch)	40 (ch,p,a)	60 (sh,ch,io,m,s)
Texaco 1940-42	N.R.	N.R.	N.R.	N.R.	N.R. (io)	N.R. (ch,p,mn,e)
UNOCAL 1962-63	0	0	0	50 (merger)	50 (merger)	50 (merger)

a = agricultural; c = coal; ch = chemical; cn = construction; e = equipment; f = fertilizer; io = international oil; m = minerals; merger = horizontal merger; mn = manufacturing; p = packaging; pl = plastic; r = real estate; s = services; sh = shipbuilding; u = uranium; N.R. = not reported; and TRANS = transitional years.

[1] Number is percent diversified sales. Markets other than domestic oil and gas are in parentheses.

[2] There is no transition period; the year is the first year of multidivisional period.

TABLE 7.5–LONG-RUN PROFITABILITY AS A FUNCTION OF ORGANIZATIONAL FORM IN CROSS-SECTION WITH OUTLIER CORRECTION

Year	DOMVI	FORVI	FORIN	FFORM	TRANS	DECNTRL	MLDV	DW	R^2
1950	0.09** (2.67)	-0.02 (-0.76)	0.12 (0.20)	4.23*** (22.91)	0.27** (2.06)	-0.16** (-2.10)	-0.18*** (-3.26)	2.66	0.68
1951	0.07+ (2.67)	-0.11** (-2.13)	0.175* (1.98)	4.46*** (33.45)	0.10 (0.65)	-0.24** (-2.51)	0.13 (0.77)	2.10	0.45
1952	0.75*** (2.34)	-0.14** (-2.57)	0.20** (2.40)	4.52*** (29.13)	0.14 (0.85)	-0.27** (-2.65)	0.12 (0.60)	2.21	0.47
1953	0.12*** (6.22)	-0.15 (-2.65)	0.19** (2.33)	4.26*** (33.48)	0.09 (0.56)	-0.27** (-2.57)	-0.24 (-0.14)	2.54	0.84
1954	0.14*** (4.17)	-0.15 (-3.75)	0.19*** (3.21)	4.10*** (22.53)	0.28*** (2.75)	-0.14+ (-1.58)	0.16+ (1.22)	2.69	0.77
1955	0.10*** (4.37)	-0.21*** (-5.22)	0.32*** (5.25)	4.42*** (41.47)	0.171 (1.32)	-0.28*** (-3.67)	0.33*** (3.24)	2.22	0.76
1956	0.08* (3.12)	-0.17** (-2.49)	0.24** (2.48)	4.59*** (36.95)	0.07 (0.53)	-0.12* (-1.33)	0.23* (1.72)	2.33	0.54
1957	0.11*** (3.65)	-0.20 (-2.37)	0.30 (2.56)	4.40*** (37.80)	0.16 (0.10)	-0.26* (-1.96)	0.05 (0.30)	2.30	0.46
1958	0.12*** (4.30)	-0.12* (-1.75)	0.18* (1.75)	4.29*** (39.25)	-0.12 (-0.56)	-0.25** (-2.01)	0.03 (0.22)	1.92	0.52
1959	0.12*** (3.94)	-0.00 (-0.08)	0.01 (0.11)	4.29*** (38.48)	-0.18+ (-1.33)	-0.14 (-1.06)	0.17 (1.15)	2.17	0.48
1960	0.07** (2.02)	0.04 (0.63)	-0.09 (-1.02)	4.31*** (31.91)	-0.11 (-0.81)	-0.01 (-0.07)	0.18 (0.24)	2.76	0.09
1961	0.07** (2.05)	-0.4 (-0.76)	-0.00 (-0.05)	4.51*** (34.27)	-0.19 (-1.19)	-0.23+ (-1.33)	0.15 (1.09)	2.47	0.15
1962	0.05 (1.18)	0.01 (0.19)	0.00 (0.03)	4.42*** (25.47)	-0.07 (-0.46)	-0.02 (-0.09)	-0.04 (-0.29)	2.59	-0.18

Year								DW	R²
1963	0.08*** (2.67)	-0.07** (-2.70)	0.10*** (2.60)	4.43*** (29.10)	-0.15 (-0.93)	-0.13 (-0.59)	-0.04 (-0.37)	2.26	0.32
1964	0.06** (2.29)	-0.08* (-1.77)	0.14** (2.08)	4.58*** (32.69)	-0.39** (-2.41)	-0.26+ (-1.45)	-0.05 (-0.49)	2.61	0.32
1965	0.02 (0.82)	-0.08 (-1.41)	0.15+ (1.66)	4.79*** (42.37)	-0.32** (-2.26)	-0.33* (-1.71)	-0.09 (-0.70)	2.42	0.14
1966	-0.14 (-0.48)	-0.07 (1.55)	0.09+ (1.43)	4.97*** (34.21)	-0.18 (-0.97)	-0.43** (-2.03)	-0.09 (-0.66)	2.16	0.11
1967	-0.00 (-0.01)	-0.03* (-1.93)	0.03+ (1.33)	4.94*** (35.33)	0.04 (0.17)	-0.39** (-2.18)	-0.14+ (-1.36)	2.07	0.27
1968	-0.01 (-0.25)	-0.03 (-0.85)	0.05 (0.93)	5.16*** (41.16)	-0.17 (-0.96)	-0.49*** (-3.13)	-0.31*** (-2.66)	1.69	0.24
1969	0.01 (0.39)	-0.06* (-1.78)	0.07+ (1.57)	4.87*** (44.55)	-0.11 (-0.70)	-0.36** (-2.00)	-0.21* (-1.83)	2.06	0.13
1970	0.00 (0.30)	-0.01 (-0.29)	0.03 (0.70)	4.67*** (51.24)	-0.42** (-2.64)	-0.42*** (-2.65)	-0.31*** (-3.01)	1.72	0.20
1971	0.02+ (1.67)	-0.07** (-2.45)	0.11*** (2.77)	4.69*** (65.44)	-0.61*** (-3.77)	-0.37*** (-2.53)	-0.37*** (-4.33)	2.06	0.50
1972	0.05*** (2.67)	-0.04 (-1.22)	0.05 (1.15)	4.72*** (83.44)	0.00 (0.00)	-0.17+ (-1.63)	-0.47*** (-4.98)	1.73	0.66

t-statistics in parentheses. Dependent variable: Log Tobin's Q; DOMVI = Log (domestic production/domestic refined); FORVI = Log (foreign production/foreign refined); FORIN = Log (foreign production/total production); FFORM = functional; TRANS = transitional; DECENTRL = decentralized forms; MLDV = multidivisional; R² = adjusted R²; and DW = Durbin-Watson statistic;

+ 20% level of significance. * 10% level of significance. ** 5% level of significance. *** 1% level of significance.

TABLE 7.6–LONG-RUN PROFITABILITY AS A FUNCTION OF ORGANIZATIONAL FORM IN CROSS-SECTION WITHOUT OUTLIER CORRECTION

Year	DOMVI	FORVI	FORIN	FFORM	TRANS	DECNTRL	MLDV	DW	R²
1950	0.09* (1.785)	0.16 (0.46)	-0.04 (-0.83)	4.23*** (16.87)	0.28* (1.71)	-0.17 (-1.25)	-0.17 (-0.76)	2.66	0.35
1951	0.11* (1.96)	0.60† (1.39)	-0.06 (-1.03)	4.29*** (16.69)	0.15 (0.61)	-0.23* (-1.53)	0.17 (0.92)	2.05	0.30
1952	0.11* (1.84)	0.71* (1.46)	-0.08 (-1.17)	4.33*** (13.69)	0.19 (0.69)	-0.24† (-1.46)	0.15 (0.23)	2.05	0.28
1953	0.19*** (3.77)	-0.00 (-1.00)	-0.01 (-0.17)	3.82*** (14.24)	0.22 (0.87)	-0.08 (-0.57)	0.24 (0.96)	2.52	0.47
1954	0.16*** (3.57)	-0.00 (-0.71)	-0.01 (-0.39)	3.95*** (16.55)	0.36** (2.47)	-0.03 (-0.25)	0.32† (1.36)	2.76	0.51
1955	0.13*** (3.76)	-0.00 (-0.68)	0.01 (0.12)	4.27*** (25.04)	0.17 (0.94)	-0.13 (-1.03)	0.39* (1.56)	2.23	0.47
1956	0.11*** (2.42)	0.03 (0.08)	-0.00 (-0.07)	4.42*** (21.12)	0.11 (0.55)	-0.01 (-0.07)	0.30† (1.59)	2.40	0.16
1957	0.14*** (3.35)	0.02 (0.04)	0.03 (0.55)	4.23*** (23.10)	0.03 (0.14)	-0.16 (-0.87)	0.14 (0.70)	2.04	0.31
1958	0.13*** (3.57)	0.26 (0.72)	-0.01 (-0.23)	4.23*** (26.56)	-0.11 (-0.49)	-0.23 (-1.36)	0.09 (0.51)	2.03	0.32
1959	0.12*** (3.34)	0.24 (0.83)	-0.03 (-0.61)	4.29*** (28.79)	0.19 (-1.16)	-0.20 (-1.25)	0.17 (1.00)	2.05	0.34
1960	0.08** (2.12)	0.42† (1.30)	-0.09* (-1.68)	4.31*** (26.18)	-0.17 (-0.96)	-0.13 (-0.68)	0.13 (0.76)	2.45	0.07
1961	0.07† (1.80)	0.55* (1.78)	-0.13** (-2.29)	4.55*** (26.62)	-0.25† (-1.42)	-0.24 (-1.18)	0.06 (0.51)	2.17	0.10
1962	0.04 (0.92)	0.22 (0.74)	-0.02 (-0.35)	4.49*** (23.50)	-0.12 (-0.58)	0.01 (0.02)	-0.06 (-0.39)	2.28	-0.21

Year								DW	R²
1963	0.07* (1.79)	0.47* (1.72)	-0.03 (-0.77)	4.44*** (27.10)	-0.19 (-0.90)	-0.04 (-0.18)	-0.04 (-0.31)	2.29	0.03
1964	0.06* (1.67)	0.12 (0.78)	0.03 (0.74)	4.58*** (27.71)	-0.36** (-1.73)	-0.28 (-1.20)	-0.08 (-0.59)	2.62	0.08
1965	0.03 (0.72)	0.25 (0.98)	0.01 (0.14)	4.74*** (28.57)	-0.34+ (-1.58)	-0.30 (-1.27)	-0.09 (-0.59)	2.45	-0.02
1966	-0.01 (-0.36)	0.16 (0.59)	-0.01 (-0.39)	4.93*** (27.56)	-0.16 (-0.59)	-0.42* (-1.63)	-0.13 (-0.79)	2.24	0.07
1967	-0.02 (-0.67)	-0.02 (-0.17)	-0.01 (-0.17)	5.07*** (30.72)	-0.04 (-0.13)	-0.47** (-2.15)	-0.18* (-1.37)	2.25	0.04
1968	-0.02 (-0.67)	-0.11 (-0.74)	0.03 (0.75)	5.27*** (30.29)	-0.30* (-1.36)	-0.61** (-2.51)	-0.37** (-2.26)	1.82	0.12
1969	-0.00 (-0.11)	-0.00 (-0.00)	0.00 (-0.02)	4.93*** (35.52)	-0.18 (-0.79)	-0.44* (-1.76)	-0.25* (-1.46)	2.12	-0.03
1970	-0.00 (-0.02)	-0.02 (-0.16)	0.02 (0.56)	4.63*** (39.69)	-0.02* (-1.45)	-0.37* (-1.83)	-0.23* (-1.65)	1.69	-0.04
1971	0.02 (0.95)	0.00 (0.00)	0.02 (0.635)	4.69*** (42.31)	-0.58** (-2.06)	-0.40* (-2.11)	-0.36** (-2.75)	2.03	0.11
1972	0.05* (1.84)	0.07 (0.65)	-0.02 (-0.54)	4.77*** (38.05)	0.00 (0.00)	-0.20 (-0.70)	-0.48*** (-3.31)	1.60	0.28

t-statistics in parentheses; Dependent variable:Log Tobin's Q; DOMVI=Log (domestic production/domestic refined); FORVI = Log (foreign production/foreign refining); FORIN=Log (foreign production/total production); FFORM= functional; TRANS=transitional;DECENTRL=decentralized forms; MLDV =multidivisional; R² = adjusted R²; DW = Durbin-Watson statistic; *20% level of significance. *10% level of significance. **5% level of significance. ***1% level of significance.

TABLE 7.7–LONG-RUN PROFITABILITY AS A FUNCTION OF ORGANIZATIONAL FORM IN CROSS-SECTION, WITH OUTLIER REMOVAL

Year	DOMVI	FORVI	FORIN	FFORM	TRANS	DCENTRL	MLDV	BIAS	DW	R²
1951	0.07+ (1.52)	0.76** (2.00)	-0.09+ (-1.67)	4.50*** (17.23)	0.06 (0.28)	-0.32** (-2.37)	0.09 (0.58)	Neg.	2.9	.47
1952	0.11* (1.84)	0.70+ (1.50)	-0.08 (-1.17)	4.32*** (13.69)	0.19 (0.69)	-0.24+ (-1.46)	0.15 (0.23)	None	2.0	.28
1953	0.19*** (3.77)	-0.00 (-1.00)	-0.01 (-0.18)	3.82*** (14.24)	0.22 (0.87)	-0.76 (-0.57)	0.24 (0.96)	None	2.5	.47
1954	0.19*** (4.56)	0.00 (0.79)	-0.00 (-0.11)	3.78*** (17.24)	0.41*** (3.14)	0.03 (0.30)	0.37* (1.82)	Pos.	2.3	.63
1955	0.13*** (2.96)	-0.00 (-0.96)	0.04+ (1.34)	4.22*** (20.14)	0.22+ (1.62)	-0.12 (-1.29)	0.44** (2.41)	Pos.	2.1	.73
1956	0.15* (1.97)	-0.39 (-0.11)	0.11 (0.22)	4.17*** (13.20)	0.21 (1.22)	0.10 (0.59)	0.39** (2.34)	Pos.	1.9	.39
1957	0.08* (1.71)	-0.20 (-0.49)	0.07 (1.23)	4.40** (24.94)	0.10 (0.53)	-0.03 (-0.19)	0.26+ (1.34)	Pos.	1.6	.45
1958	0.08* (1.99)	0.47+ (1.58)	-0.06 (-1.22)	4.31** (26.90)	0.02 (0.12)	-0.09 (-0.70)	0.29** (2.16)	Pos.	1.5	.64
1959	0.08* (1.99)	0.15 (0.70)	0.00 (0.06)	4.30*** (28.55)	-0.23+ (-1.56)	-0.06 (-0.45)	0.26* (1.95)	Pos.	1.9	.65
1960	0.09** (2.50)	0.49+ (1.54)	-0.07+ (-1.32)	4.28*** (29.17)	-0.16 (-1.03)	-0.16 (-0.97)	0.21+ (1.36)	Neg.	2.2	.29
1961	0.11*** (2.74)	0.79*** (2.73)	0.14** (2.86)	4.37*** (24.73)	-0.27+ (-1.70)	-0.20 (-1.16)	0.14 (0.93)	Neg.	1.9	.34
1962	0.09*** (2.68)	0.02 (0.07)	0.01 (0.24)	4.24*** (25.03)	-0.02 (-0.61)	-0.00 (-0.02)	-0.50 (-0.40)	None	2.7	.36
1963	0.13*** (3.66)	0.04* (1.76)	-0.00 (-0.16)	4.13*** (26.10)	-0.12 (-0.71)	0.00 (0.01)	0.21 (0.19)	Pos.	2.6	.40
1964	0.06+ (1.62)	0.06 (0.46)	0.01 (0.48)	4.67*** (30.82)	-0.37* (-1.97)	-0.32* (-1.54)	-0.14 (-1.05)	Neg.	2.6	.28

1965	0.02 (0.52)	0.25 (1.08)	0.01 (0.19)	4.85*** (31.43)	-0.42** (-2.16)	-0.38* (-1.78)	-0.16 (-1.15)	Neg.	2.5	.18
1966	-0.03 (-1.02)	0.16 (0.70)	0.01 (0.25)	5.08*** (33.93)	-0.29+ (-1.33)	-0.55*** (-2.64)	-0.31** (-2.21)	Neg.	2.4	.31
1967	-0.03 (-1.04)	-0.02 (-0.22)	-0.03 (-0.97)	5.19*** (37.24)	-0.15 (-0.65)	-0.53*** (-3.04)	-0.20* (-1.75)	Neg.	2.3	.39
1968	0.01 (0.24)	-0.09 (-0.79)	0.02 (0.71)	5.12*** (23.69)	-0.27 (-0.29)	-0.55** (-2.42)	-0.35** (-2.12)	None	1.5	.38
1969	0.03 (1.28)	0.01 (0.11)	-0.00 (0.16)	4.80*** (42.21)	-0.18 (-1.11)	-0.41** (-2.41)	-0.26** (-2.12)	None	1.9	.54
1970	0.00 (0.30)	-0.02 (-0.28)	0.03+ (1.58)	4.79*** (65.00)	-0.58*** (-4.00)	-0.57*** (-4.59)	-0.47*** (-5.09)	Neg.	1.9	.64
1971	0.02+ (1.51)	-0.04 (-0.82)	0.04** (2.61)	4.71*** (83.73)	-0.65*** (-4.54)	-0.43*** (-4.52)	-0.39*** (-5.46)	None	2.4	.77
1972	0.05*** (3.11)	0.04 (0.72)	0.02 (1.14)	4.77*** (62.44)	0.00 (0.00)	-0.34** (-2.00)	-0.59*** (-6.53)	None	1.4	.74

Note: t-statistics in parenthesis.

Dependent variable: Log Tobin's Q; DOMVI = Log (domestic production/domestic refined); FORVI = Log (foreign production/foreign refining); FORIN = Log (foreign production/total production); FFORM = functional; TRANS = transitional; DECENTRL = decentralized; and MLDV = multidivisional.

* 10% level of significance. ** 5% level of significance. *** 1% level of significance. + 20% level of significance.

TABLE 7.8–PRODUCT LINE AND PRODUCTION CHARACTERISTICS COMPARISON OF ALTERNATIVE ORGANIZATIONAL GROUPINGS[1]

Study Period	Organizational Form	Vertical[2] Integration	Petrochemical, Uranium	International Oil	Number of Firms
1955-59	MLDV	100	80	66	6
1955-59	DECNTRL	40	80	60	5
1955-59	FNCTNL	33	0	0	10
1960-64	MLDV	84	75	60	12
1960-64	DECNTRL	0	66	0	3
1960-64	FNCTNL	50	8	0	12
1965-68	MLDV	74	85	66	16
1965-68	DECNTRL	0	50	50	2
1965-68	FNCTNL	11	22	0	9
1969-72	MLDV	80	85	62	19
1969-72	DECNTRL	0	50	0	2
1969-72	FNCTNL	40	0	0	5

MLDV = multidivisional firms; DECNTRL = decentralized firms; and FNCTNL = functionally organized firms.

[1] Percent of firms engaged in business activity.

[2] A firm was assumed to be vertically integrated if its self-sufficiency ratio was greater than 50%.

TABLE 7.9–LONG-RUN PROFITABILITY AS A FUNCTION OF AVERAGE LONG-RUN
PROFITABLILITY OF CONTROL FIRMS AND ORGANIZATIONAL FORM[1]

Firm	Year[2]	Portfolio	Structure	AVGQ	FFORM	TRANS	MLDV	R^2	DW[3]
Belco	1969	A	Functional	0.23 (.63)	0.61** (2.41)	0.00	-0.39*** (-2.82)	0.51	1.82
Getty Oil	1959	B	C.M.	1.29*** (3.17)	-0.15 (-0.63)	0.00	-0.19 (-1.34)	0.44	2.29
Kerr-McGee	1955	C	Functional	0.86** (2.08)	0.23 (0.72)	-0.20 (-0.88)	-0.20 (-1.44)	0.16	1.32
CONOCO	1951	E	Functional	1.49*** (4.26)	-0.19 (-0.74)	-0.15 (1.05)	0.09 (0.67)	0.77	1.50
Marathon Oil	1963	B	Functional	0.53*** (4.59)	0.26* (1.96)	0.03 (0.40)	0.15** (2.43)	0.46	2.29
UNOCAL	1964	B	Functional	0.49*** (3.06)	0.00 (0.01)	0.14 (1.59)	0.08 (1.27)	0.27	2.44
Tenneco	1962	B	Functional	0.25*** (4.35)	0.59*** (5.96)	0.00 (0.08)	-0.13*** (-4.23)	0.86	2.30
Texaco	1943	F	Functional	0.86*** (7.76)	0.16** (2.02)	0.07 (1.36)	-0.15*** (-3.67)	0.78	1.58
Ashland Oil	1970	A	Functional	0.40** (2.29)	0.25*** (2.93)	-0.00 (-0.06)	-0.18*** (-2.73)	0.52	1.79
ARCO	1966	A	Functional	0.24 (.82)	0.44 (0.40)	-0.02 (-0.09)	0.39*** (2.91)	0.40	1.90
Sun Oil Co.	1971	A	Functional	0.40*** (3.24)	0.33*** (2.72)	-0.11* (-1.73)	-0.13** (-2.27)	0.72	1.79
Occidental	1972	A	Functional	0.28 (0.69)	0.54 (1.57)	-0.15 (-0.86)	-0.33* (-1.82)	0.35	2.12
Gulf Oil	1958	C	Functional	0.52*** (3.24)	0.09 (1.13)	0.14* (1.80)	0.07 (1.59)	0.38	1.73
Murphy	1972	A	F.S.	0.47*** (3.70)	0.24 (1.57)	0.00	0.05 (0.98)	0.50	1.52
St. of Ohio	1963	B	F.S.	0.69*** (4.11)	-0.09 (-1.11)	0.10 (1.02)	0.17** (2.79)	0.40	1.51

Amoco	1961	B	F.S.	0.42***	0.035	-0.04	0.09***	0.46	2.07
				(4.61)	(0.85)	(-0.93)	(2.48)		
Shell Oil	1961	D	F.S.	0.84***	0.09	0.08	0.05	0.45	1.80
				(3.45)	(0.71)	(0.98)	(0.64)		
Cities Service	1967	B	Hold. Co.	0.26**	0.09	0.10*	0.04	0.13	1.43
				(2.02)	(1.10)	(1.81)	(0.79)		
Chevron	1955	C	F.S.	0.55***	0.21	0.10	0.12***	0.74	1.86
				(4.93)	(0.07)	(1.21)	(3.21)		
Mobil Corp.	1960	B	F.S.	0.50***	-0.01	-0.06	0.11**	0.79	1.88
				(8.65)	(-0.44)	(-1.34)	(5.21)		
Phillips	1975	A	Hold. Co.	0.36**	0.30***	0.27***	0.08	0.43	2.05
				(2.65)	(2.85)	(3.27)	(1.61)		
Exxon	1966	B	F.S.	0.84***	-0.03	0.10**	0.01	0.77	2.29
				(8.21)	(-0.59)	(2.63)	(0.30)		

Portfolio A = American Petrofina, Commonwealth, Superior, Texaco.
Portfolio B = Superior, Texaco, Phillips, CONOCO.
Portfolio C = Sun, Sunray, Superior, Phillips.
Portfolio D = Ashland, Phillips, Sun, Sunray.
Portfolio E = Ashland, ARCO, Superior, Marathon Oil, UNOCAL, Sun, Sunray.
Portfolio F = Ashland, ARCO, Getty, Marathon Oil, UNOCAL, Sunray, Sun.
t-statistics in parentheses; Dependent variable=Tobin's Q; C.M.=corrupted multidivisional form; AVGQ = average Tobin's Q for a portfolio of firms; FFORM=functional; TRANS=transitional; MLDV = multidivisional; DW = Durbin-Watson statistic; and R^2 = adjusted R^2.
* 10% level of significance. ** 5% level of significance. *** 1% level of significance.
[1] Firms ranked by size within centralized and decentralized groups. [2] Year is first year of the multidivisional period. [3] Durbin-Watson test rejects autocorrelated error for all firms except Kerr-McGee.
Note: Dashed line separates centralized from decentralized firms.

TABLE 7.10–CHANGE IN RELATIVE TOBIN'S Q OVER TIME
FOR ALTERNATIVE PORTFOLIOS[1]

Years before Transition	Years after Transition	Portfolio	Sample Size	Difference of Means	t-Statistic
-10	10	all firms	22	-0.084	-0.539
-5	10	all firms	22	0.058	0.475
-10	10	all firms trans before 65	15	-0.022	-0.126
-5	10	all firms trans before 65	15	0.133	1.000
-10	10	all functionally organized firms[2]	13	-0.178	-0.835
-5	10	all functionally organized firms[2]	13	0.037	0.253
-10	10	functionally organized firms transition before 65	9	-0.148	-0.568

-5	10	functionally organized firms transition before 65	9	0.144	0.616
-10	10	all decentralized firms	9	0.153	1.224
-5	10	all decentralized firms	9	0.213	1.260
-10	10	eight largest decentralized firms	8	0.168	1.200
-5	10	eight largest decentralized firms	8	0.191	1.080
-10	10	eight largest functional firms	8	-0.008	-0.030
-5	10	eight largest functional firms	8	0.105	0.724

Note: Firms changing organizational form before the OPEC period of 1973 and after 1963 have a final multidivisional year at 1972, yielding a multidivisional period of less than ten years. If firms changed after 1973, their final multidivisional period was assumed to stop at the second oil shock in 1979.
[1] Means comparison.
[2] Occidental was excluded because it was expelled from Libya in the test year.

TABLE 7.11–MEANS COMPARISON OF ALTERNATIVE PORTFOLIOS

Time Period	Control Firms	FNCTNL Firms	MLDV Firms	DECNTRL Firms	Difference of Means	t-Statistic
1956-72	1.40 (0.66)	-	1.16 (0.36)	-	-0.21	-1.88**
1951-72	-	1.21 (0.42)	1.17 (0.37)	-	-0.04	-0.661
1951-72	-	-	1.17 (0.37)	1.00 (0.23)	0.17	3.07***

Standard deviation in parenthesis; Control Firms = portfolio of Tobin's Q's for firms that do not change organizational form. FNCTNL = portfolio of Tobin's Q's for firms that are functionally organized with 10 percent or less in diversified sales. MLDV = portfolio of Tobin's Q's for multidivisional firms. DECNTRL= firms with either a holding company form or functional-with-subsidiary form.
*** 1% of significance.　** 5% of significance.　* 10% level of significance.

TABLE 7.12.-AVERAGE TOBIN'S Q OF A PORTFOLIO OF MULTIDIVISIONAL FIRMS AS A FUNCTION OF AVERAGE TOBIN'S Q OF PORTFOLIOS OF OTHER FIRMS

Time Period	Dependent Variable	Intercept	CNTRL	FNC	DECNTRL	Corrected Lags	R^2
1956-72	PRTMLDV	-0.03 (-0.21)	0.91*** (8.43)	-	-	2	0.90
1951-72	PRTMLDV	1.15*** (6.54)	-	0.10 (0.88)	-	2	0.55
1951-72	PRTMLDV	-0.10*** (3.83)	-	-	1.41 (0.88)	2	0.85

Note: t-statistics in parenthesis.
PRTMLDV = portfolio of Tobin's Q for multidivisional firms; CNTRL = portfolio of control firms; FNC= portfolio of functionally organized firms; DECNTRL = portfolio of decentralized firms, either a holding company or F.S. form; CORRECTED LAGS = number of periods adjusted for autocorrelated error; and R^2 = adjusted R^2.
*** = 1% level of significance. ** = 5% level of significance.

TABLE 7.13–CHANGE IN EARNINGS AS A FUNCTION OF CHANGES IN SALES, PRICES, AND ORGANIZATIONAL FORM[1]

Firm	Year[2]	Structure	DSALES	DOILP	FFORM	TRANS	MLDV	DW[++]	R²
Belco	1969	Functional	0.63 (0.91)	-0.06 (-0.03)	-0.00 (-0.01)	0.00	0.06 (0.15)	2.01	-0.13
Getty Oil	1959	C.M.	0.96*** (20.4)	0.18 (0.23)	-0.01 (-0.19)	0.00	-0.03 (-0.57)	2.59	0.95
Kerr-McGee	1955	Functional	1.17 (0.77)	2.57 (0.39)	-0.15 (-0.15)	-0.02 (-0.02)	0.14 (0.14)	2.63	-0.26
CONOCO	1951	Functional	1.15*** (2.77)	0.61 (0.74)	0.07 (1.11)	-0.11 (-1.03)	-0.07 (-1.09)	1.54	0.30
Marathon Oil	1963	Functional	0.58** (2.40)	0.39 (1.37)	-0.02 (-0.61)	-0.03 (-0.64)	0.07** (2.12)	2.05	0.47
UNOCAL	1964	Functional	2.65*** (6.64)	-0.23 (-0.33)	-0.00 (-0.03)	0.03 (0.11)	-0.25* (-1.60)	1.95	0.65
Tenneco	1962	Functional	1.31* (2.16)	0.16 (0.14)	-0.12 (-0.73)	0.06 (0.28)	0.05 (0.33)	2.57	0.06
Texaco	1943	Functional	3.41*** (3.05)	1.90 (0.08)	-0.22 (-0.71)	0.23 (0.39)	0.07 (0.17)	2.72	0.19
Ashland Oil	1970	Functional	7.59*** (3.37)	-4.04** (-1.95)	-0.50 (-1.25)	-0.58 (-0.78)	0.52 (0.87)	2.53	0.30
ARCO	1966	Functional	1.61* (9.93)	-0.52*** (-2.76)	0.01 (0.29)	-0.11 (-1.06)	-0.04 (-0.67)	2.44	0.83
Sun Oil Co.	1971	Functional	0.07* (2.12)	0.69** (2.26)	0.06* (1.34)	-0.11 (-1.26)	-0.21*** (-2.77)	1.57	0.52
Occidental	1972	Functional	1.16*** (4.44)	0.15 (0.09)	-0.37 (-1.04)	0.63* (1.38)	0.33 (0.59)	2.27	0.59

	1958	Functional							
Gulf Oil			2.08*** (8.11)	0.04 (0.07)	-0.08** (-2.33)	-0.10+ (-1.33)	0.07+ (1.81)	2.29	0.75
Murphy	1972	F.S.	1.34 (1.00)	0.08 (0.04)	-0.25 (-0.81)	0.00	0.49 (0.81)	2.38	-0.02
St. of Ohio	1963	F.S.	0.49 (0.87)	0.35 (0.95)	-0.02 (-0.32)	0.13 (0.91)	0.00 (0.00)	1.51	-0.08
Amoco	1961	F.S.	3.15*** (5.24)	0.12 (0.11)	-0.17** (-2.46)	0.14+ (1.42)	0.12+ (1.51)	2.96	0.60
Shell Oil	1961	F.S.	2.74*** (7.42)	0.01 (0.01)	-0.18*** (-2.64)	0.06 (0.77)	0.05 (0.53)	2.23	0.70
Cities Service	1967	Hold. Co.	1.61*** (3.35)	0.00 (0.00)	-0.04 (-1.10)	0.10+ (1.58)	0.01 (0.19)	1.67	0.46
Chevron	1955	F.S.	1.75*** (4.10)	-0.40 (-0.38)	-0.06 (-1.19)	-0.06 (-0.41)	0.01 (0.18)	1.98	0.45
Mobil Corp.	1960	F.S.	1.32*** (6.11)	0.20 (0.30)	-0.06** (-2.10)	0.04 (0.38)	0.06+ (1.48)	1.74	0.66
Phillips	1975	Hold. Co.	1.29*** (3.63)	-0.03 (-0.11)	-0.04+ (-1.42)	0.14+ (1.53)	0.10+ (1.75)	1.86	0.83
Exxon	1966	F.S.	1.12+ (1.70)	-0.17 (-0.28)	-0.06 (-0.61)	0.21+ (1.59)	0.05 (0.41)	2.10	0.10

Note: t-statistics in parenthesis.
C.M. = corrupted multidivisional form; Dependent variable = first difference of log of earnings; DSALES = first difference of log of sales; DOILP = first difference of log of oil prices; FFORM = functional form; TRANS = transitional form; MLDV = multidivisional form; DW = Durbin-Watson statistic; and R^2 = adjusted R^2.
[1] Firms ranked by size at the time of transition within centralized and decentralized categories. [2] Year equals first year with the multidivisional form. + 20% level of significance. ** 5% level of significance. *** 1% level of significance. ++ Durbin-Watson test rejects autocorrelated error for all firms.
Note: Firms above the dashed line are centralized and those below are decentralized.

TABLE 7.14—Change in General and Administrative Expenses as a Function of the Changes in Refined Barrels of Oil and Organizational Form[1]

Firm	Year[2]	Structure	DOIL	FFORM	TRANS	MLDV	DW	R²
Belco	1969	Functional	0.03 (0.05)	0.25 (1.15)	0.00	-0.17 (-0.65)	1.84	-0.09
Getty	1959	C.M.	0.80*** (14.2)	0.91+ (1.60)	0.00	-0.49 (-0.53)	2.05	0.91
CONOCO	1951	Functional	0.30 (1.35)	-0.03 (-0.88)	0.07 (1.26)	0.08*** (2.16)	1.50	0.11
Marathon Oil	1963	Functional	0.29* (1.84)	0.33 (0.64)	-0.01 (-0.11)	-0.31 (-0.46)	2.38	0.04
UNOCAL	1964	Functional	1.14*** (12.07)	0.01 (0.58)	0.00 (0.00)	-0.04 (-1.20)	0.92	0.87
Tenneco	1962	Functional	1.56*** (2.76)	0.30 (1.17)	-0.47* (-1.62)	-0.14 (-0.42)	2.04	0.59
Texaco	1943	Functional	0.12 (0.39)	0.00 (0.01)	-0.04 (-0.58)	0.18 (0.35)	1.50	-0.10
Ashland	1970	Functional	0.73*** (4.37)	0.06*** (4.34)	0.07* (2.15)	-0.07*** (-3.90)	1.96	0.79
ARCO	1966	Functional	0.88*** (9.96)	-0.01 (-0.37)	-0.02 (-0.31)	0.01 (0.45)	1.25	0.85
Sun Oil Co.	1971	Functional	0.67*** (2.70)	0.01 (0.29)	0.09 (1.18)	0.03 (0.25)	1.94	0.19
Occidental	1972	Functional	-0.44** (-1.61)	1.00*** (3.65)	-0.86*** (-2.77)	-0.90*** (-3.23)	2.27	0.59
Gulf	1958	Functional	0.68** (2.48)	0.01 (0.21)	0.16*** (3.41)	-0.00 (-0.90)	1.43	0.56

Murphy	1972	F.S.	0.21 (1.13)	0.10** (1.96)	0.00	-0.07 (-0.94)	2.11	0.07
Amoco	1961	F.S.	0.16 (0.06)	0.04 (1.28)	-0.03 (-0.80)	-0.00 (-0.16)	2.38	-0.12
Shell Oil	1961	F.S.	0.53** (2.41)	-0.00 (-0.04)	0.02 (0.63)	0.02 (0.47)	2.75	0.15
Cities Service	1967	Hold. Co.	0.04 (1.21)	0.18** (2.10)	-0.62* (-1.76)	-0.07** (-2.29)	2.57	0.24
Mobil Corp.	1960	F.S.	0.07+ (1.63)	0.05 (0.14)	0.03 (0.21)	-0.01 (-0.20)	1.54	-0.14
Phillips	1975	Hold. Co.	0.25 (0.59)	0.04 (1.16)	-0.08 (-1.01)	-0.08* (-1.90)	1.98	0.14
Exxon	1966	F.S.	0.16*** (3.06)	0.11 (0.59)	-0.05 (-1.02)	-0.08** (-1.92)	2.40	0.09

C.M. = corrupted multidivisional form; Dependent variable = log of first difference of general and administrative costs. DOIL = first difference of log of refined barrels of oil; FFORM = functional form; TRANS = transitional form; MLDV = multidivisional form; DW = Durbin-Watson statistic; and R_2 = adjusted R^2.

Note: Firms ranked by size at the time of transition within centralized and decentralized categories.

+ 20% level of significance. * 10% level of significance. ** 5% level of significance. *** 1% level of significance.

[1] Firms ranked by size at the time of transition within centralized and decentralized categories. t-statistics in parenthesis.

[2] Year equals first year with the multidivisional form.

Note: Centralized firms are above the dashed line and decentralized fall below it.

TABLE 7.15–Change in Diversified Sales as a Function of Its Lagged Change and Organizational Form–Functional Firms and Decentralized Firms with Less than 10 Percent in Sales

Firm	Year[1]	Diverse	Structure	DDVRSFD	FFORM	PRIOR	TRANS	MLDV	DW	R^2
Belco	1969	20	Functional	0.19 (0.80)	-0.10 (-0.02)	1.59** (1.98)	0.00	-0.08 (-0.13)	2.35	0.14
Murphy	1972	5	C.M.	0.00 (0.27)	0.03 (0.02)	0.63*** (2.96)	0.00	0.11 (0.87)	2.44	0.27
CONOCO	1951	4	Functional	-0.06 (-0.52)	-0.05 (-0.40)	0.02 (0.12)	0.19 (0.96)	0.29** (2.04)	1.97	0.04
Marathon Oil	1963	0	Functional	-0.12 (-0.47)	-0.02 (-0.05)	-0.00 (-0.01)	0.01 (0.01)	0.81* (1.38)	2.07	-0.08
UNOCAL	1964	0	Functional	0.01 (0.07)	0.06 (1.01)	0.03 (-0.33)	0.40*** (3.37)	-0.02 (-0.22)	2.83	0.31
St. of Ohio	1963	3	Functional	-0.43** (-2.08)	-0.26 (-0.76)	2.14*** (3.20)	-1.20* (-1.66)	0.63* (1.46)	1.36	0.32
Tenneco	1962	0	Functional	-0.48** (-2.36)	-0.03 (-0.01)	-0.01 (-0.01)	3.49*** (3.94)	0.37 (0.69)	1.86	0.38
ARCO	1966	0	Functional	-0.30 (-1.26)	-0.03 (-0.45)	0.06 (0.46)	0.60* (1.83)	0.24** (2.41)	2.26	0.13
AMOCO	1961	6	Functional	-0.56 (-0.27)	0.05 (0.87)	0.02 (0.28)	0.06 (0.96)	0.19*** (2.95)	2.22	0.37

Company	Year[1]		Form							
Sun Oil Co.	1971	7	Functional	-0.26 (-1.13)	-0.11 (-0.81)	0.60*** (2.68)	0.12 (0.41)	0.38** (2.37)	1.50	0.23
Shell Oil	1961	8	F.S.	0.42* (1.94)	-0.03 (-0.36)	0.17+ (1.42)	0.11 (1.29)	0.08 (0.75)	2.46	0.07
Cities Service	1967	19	Hold. Co.	-0.02 -(0.10)	-0.09 (-0.37)	1.26*** (2.61)	0.16 (0.33)	0.10 (0.27)	2.33	0.13
Exxon	1966	46	F.S.	-0.55*** (-3.18)	-0.02 (-0.06)	3.99*** (5.24)	0.40 (0.62)	0.12 (0.25)	2.17	0.56

[1] Year of end of transition.

Dependent variable = first difference of log of change in diversified sales; C.M. = corrupted multidivisional form. DDVRSFD = first difference of log of lagged diversified sales; FFORM = functional form; TRANS = transitional form; MLDV = multidivisional form; and Diverse at Trans = percent of diversified sales at transition period.

Note: Firms ranked by size of diversified sales at the time of transition. T-statistics in parenthesis in other columns.

+ 20% level of significance. * 10% level of significance. ** 5% level of significance. *** 1% level of significance.

FIGURE 7.1-CAR FOR FUNCTIONAL FIRMS

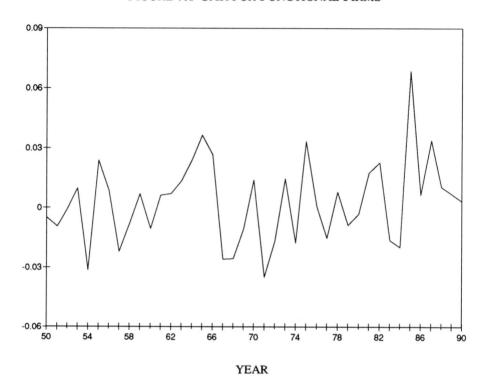

YEAR

FIGURE 7.2-CAR FOR DECENTRALIZED FIRMS

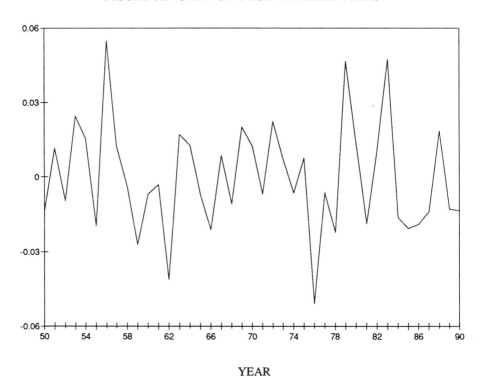

YEAR

Chapter 8

Firm Strategies and Histories 1973-90

In the oil industry, the period prior to 1973 was a time of growth into related markets as firms entered foreign oil and petrochemical markets. In response to these investments, managers adopted the multidivisional form as an efficient tool for allocating human and financial resources. Williamson (1975) claims that managers of multidivisional firms have fewer investment opportunities within the firm than do investors in an external market but have superior information than do outside investors. Thus, he asserts that the internal operations of a multidivisional firm can mimic an external capital market in efficiency. Chapter 4 suggested that diversification into highly related businesses by oil companies was advantageous. In this chapter, the ability of the multidivisional form to accommodate unrelated growth will be evaluated.

The pre-1973 period was a time of numerous opportunities for related investments into international oil, petrochemical, and coal markets. The investment climate, however, changed from the mid-1970's to the mid-1980's. First, the major oil producing countries under OPEC asserted themselves by acting in unison to periodically raise oil prices, thereby reducing company profits on foreign produced oil but raising them on domestic oil. Second, rising nationalism in many countries resulted in oil property nationalizations. Third, by 1973, the petrochemical business had reached a mature stage of its product cycle. Hence, opportunities for related investments diminished, while profits rose on domestic oil and dropped on foreign oil. Each firm reacted differently to these changed circumstances, yet many commonalities existed.

This chapter is organized such that company histories are provided for each firm discussed in Chapter 4. A discussion follows the company histories.

8.1 Company Histories

The period 1973-90 was a time of high cash flow levels for oil company managers but few related investment opportunities. Many companies undertook broad diversification strategies in order to use their cash flow. Almost all of these strategies failed.

8.1.1 *AMOCO*. Having less invested in international oil production but also having a mediocre position in domestic oil, Amoco suffered little from the nationalization of oil company assets by several oil producing countries in the early 1970's but also gained little from domestic performance after the escalation in oil prices after 1973. Initially, Chairman John Swearingen responded to the new environment by increasing domestic oil exploration, buying partial stakes in Cetus and Analog Devices, and diversifying the chemical business through internal development.

A modification of strategy occurred, however, after the second oil price hike as Amoco expanded its diversification strategy. Speaking before the price increase, Swearingen said, "Coal is dog's breakfast" (*Businessweek*, 1978a). After the change in prices, however, Swearingen bought Cyprus Mines, Emerald Mines, and Harbart Mines in 1979, thereby becoming a large coal and mineral producer.

As investor pressure grew in the early 1980's, managers changed strategy once again, but this time to horizontal growth and stock value enhancements. In 1984, Chairman Richard Morrow repurchased $2 billion worth of stock and made plans to discard many of the diversified businesses of Amoco. By 1988, he had sold the 16 percent stake in Analog Devices, spun off Cyprus Mines, and began purchasing oil properties. The biggest horizontal purchase occurred in 1988 with the acquisition of Dome Petroleum of Canada for $4.2 billion.

8.1.2 *ARCO*. Obtaining most of its oil production from domestic sources, ARCO suffered few disruptions from the oil property nationalizations by foreign governments that occurred during the 1970's. Moreover, because its Alaskan crude oil production came on stream shortly after the price increases of the early 1970's, it had a high level of cash flow that it could use for future investments.

Although Chairman Robert Anderson emphasized investment in domestic oil production, he also chose to further broaden the company by diversifying into metals. His purchase of Anaconda for $700 million at first appeared to be a solid move as the price of metals rose. Later price declines, coupled with the high cost nature of the Anaconda mines, yielded a $642 million loss from 1982 to 1984 (Ticer and Glasgall, 1985).

In the oil business, Anderson directed most of ARCO's funding to traditional domestic drilling but did undertake overseas ventures and participated in a shale oil venture with Exxon. Citing poor shale oil prospects, however, Anderson sold the ARCO holdings in 1980.

Circumstances at ARCO changed significantly after 1980 as investors began to exert pressure on management for better performance. In 1984, after aggressiveness in the capital markets had led to takeovers at Getty, Cities Service, and Gulf, Chairman William Kieschnick, who took control in 1982, bought back one-third of ARCO's stock for $4.6 billion and increased debt. Investors approved of the moves, bidding up stock prices by about 25 percent, which, combined with higher debt, made ARCO a less attractive takeover target.

Burdened with a debt load of 60 percent of capital and pressure to improve performance, Kieschnick accelerated a consolidation program begun by Anderson. Kieschnick wrote off $1.3 billion in assets in 1985 and made plans to sell Anaconda, exit from the East Coast marketing area, and reduce capital investment in oil. By the time Kieschnick and Lodwick Cook, his replacement, had finished consolidating, they had reduced the workforce by 33 percent and had restructured the company into three activities: marketing oil in the western half of the United States, producing and exploring for oil, and manufacturing chemicals. In this reduced form, Cook upgraded service stations and began to expand exploration efforts overseas.

8.1.3 *Ashland Oil.* Although Ashland Oil has traditionally been a crude-deficient oil refiner, it was little affected by the first major escalation of oil prices in 1973. At that time, Ashland purchased much of its foreign oil from a politically stable Iran and from domestic producers.[1] The combination of ample sources of oil and highly efficient refining operations enabled Ashland to outperform many of its larger vertically-integrated competitors in the mid-1970's.

The success that Ashland had in refining, however, was not matched by its production unit. Aside from a finding cost of $3.75 per barrel in 1977 versus the industry average of $3.09, Chairman Orin Atkins gave reasons such as changed tax policies, nationalizations abroad, and the high cost of developing reserves as the costs of becoming an oil producer (*Businessweek*, 1978b). As a result, Atkins sold Ashland Oil-Canada and most of Ashland's U.S. oil production business in 1978-79, using the proceeds of about $1 billion for stock repurchases, dividends, and acquisitions.

The decision to sell the production units proved disastrous for Ashland, as the price of crude oil rose from $13 to $38 per barrel by 1981. Moreover, since Ashland received about 25 percent of its oil from Iran, its refineries were threatened with inefficient operating rates when the Iran-Iraq war broke out in 1979.

---

[1]Domestic producers could not raise prices to meet world prices because of price ceilings on previously discovered oil.

Ashland was able to weather its shortfall only because government regulations required crude-rich companies to share oil with refiners. Nonetheless, Ashland did feel compelled to increase payments to Middle-Eastern middlemen, who were attempting to secure access to crude oil for Ashland. After charges of bribery were made by Ashland Vice-President of Oil Supply Bill McKay, an internal investigation exonerated Ashland management but only because of a technicality in the law (*Businessweek*, 1988).

Using the remaining cash from the sale of Ashland's oil production business and perceiving a threat to Ashland's oil supply, Chairman Atkins decided to undertake a diversification program as a way to lessen Ashland's reliance on the oil business. The strategy of unrelated growth was shortlived. The two acquisitions that Atkins made, Integon, an insurance company, and U.S. Filter, a pollution control company, cost about $640 million and failed to be profitable. Shortly after the purchases, John Hall replaced Atkins as chairman. He immediately took a $270 million write-off on the new acquisitions, sold the businesses, and repositioned Ashland in the oil business by reducing Ashland's reliance on foreign oil supplies to 30 percent through the purchase of Scurlock Oil Company, a Houston crude oil distributor. Moreover, he reemphasized oil production, devoting 50 percent of 1985 capital expenditures to oil exploration and upgrading service stations (*Forbes*, 1985).

Ashland's problems did not end in the early 1980's. In 1986, Ashland management was convicted of bid-rigging on Florida construction projects, made a $25 million settlement with the SEC over payments it had made to Iran in 1979, overcame a takeover attempt by the Belzbergs of Canada, and experienced the rupture of an oil tank on the Monongahela river that dumped 500,000 gallons of oil into the river. In spite of these problems, however, Ashland's stock value doubled from 1985 to 1988, as management eventually overcame the problems that it had created for itself in the late 1970's and early 1980's.

8.1.4 *Chevron.* With 70 percent of its production overseas, Chevron had much at risk when OPEC raised oil prices in the early 1970's. Being a partner in ARAMCO with a politically moderate Saudi Arabia, however, it retained its supply links but did suffer a loss in profits per barrel of oil produced. As prices were rising, price margins on the domestic oil that Standard produced rose significantly.

Realizing a jump in cash flow but also feeling susceptible to the risks of international oil, Chairman Harold Haynes purchased 20 percent of AMAX, a mining company, for about $400 million in 1975. After another increase in oil prices in 1979, managers of Chevron made a bid for the 80 percent of AMAX that it did not own in 1981. *Businessweek* (March 1981a) suggested that the motive of management was to use the cash flow of Chevron to develop AMAX mining assets. The deal, however, failed.

After the AMAX bid, the strategy at Chevron changed to one of horizontal growth and avoidance of takeover threats. Chairman George Keller bid $13.2 billion for Gulf in an effort to increase debt from the 35 percent of capital level that it had reached and to augment domestic oil capacity (*Businessweek*, 1984c). Over the next several years, managers at Chevron concentrated on reducing debt and selling assets that did not fit a new strategy of related growth and marketing and refining in the western United States. Hence, Keller and, later, Chairman Ken Derr sold the northeastern and southeastern domestic marketing and refining operations of Gulf, 60 percent of Gulf Canada, and the AMAX holdings. Finally, after reducing debt to 33 percent of capital, Derr devoted capital expenditures to upgrading refineries and service stations and focusing exploration efforts on the Arctic and developing countries.

8.1.5 *Cities Service*. Although Cities Service produced 100,000 barrels of domestic crude per day in 1973 and realized an increase in cash flow in the early 1970's, its managers chose not to diversify beyond their current ventures. Rather, managers chose to emphasize oil production and develop their plastics, copper, and tar-sands businesses. Investments by management, however, were not successful, as the company failed to replace the oil that it produced, while investments in copper, industrial chemicals, and plastics were plagued with losses.

After a peak in profitability in 1980 but with a market value of $46 per share and an appraised asset value of $109 per share, Boone Pickens, through Mesa Oil Company, began acquiring stock in the hope of a takeover (Santry and Ord, 1981). In response to Pickens, Chairman Charles Waedelich of Cities Service solicited a merger offer from CONOCO, which had been under takeover pressure from Joseph R. Seagrams and Sons, Inc. The merger bid, however, failed as CONOCO was taken over by DuPont in September, 1981. Another merger attempt failed in 1982 as Gulf withdrew its acquisition offer. A final merger offer from Occidental Petroleum succeeded as it bought Cities service for about $4 billion in 1982.

8.1.6 *CONOCO*. Although CONOCO had a strong domestic production unit, its profits from 1973 to 1979 barely kept pace with inflation. With the cash flow that did come from operations, Chairman Howard Blauvelt reduced debt and attempted to correct operating problems. In the coal business, for example, CONOCO reported one good year followed by several unprofitable ones as first the United Mineworkers went on strike and then a recession affected the market for metallurgical coal. CONOCO encountered other problems in their offshore exploration and domestic oil production units. Investments in the Baltimore Canyon, for example, yielded no oil, while the British raised taxes on North Sea production. In domestic oil production, Blauvelt centralized control of the sub-unit after firing several of the top managers. In spite of the difficulties at CONOCO, however, suggested that, by 1979, Blauvelt had reshaped CONOCO into a form that would thrive in the 1980's when investments in oil would pay off (Flanigan, 1979).

The increase in oil prices in 1979 and the location of CONOCO Oil production operations in riskless political locations made prospects for CONOCO bright. Yet, investors valued the stock at only $62 per share versus its $138 asset value (Flanigan, 1979). In response to this opportunity, Joseph Seagrams and Sons, Inc. announced its intention to acquire a large share of the stock in 1981. Mobil, Texaco, and DuPont followed the lead of Seagrams with offers of their own. Eventually, DuPont won the bidding contest with an $88 per share offer.[2]

8.1.7 *Exxon*. Although foreign governments nationalized many of its oil properties, Exxon, because of its position as the leading domestic oil producer, benefited greatly from the events of the 1970's. Chairman Clifton Garvin used the proceeds from the high price of oil to quadruple capital expenditures on energy and nonenergy businesses between 1973 and 1982. Investments included: $2 billion for information processing, $1.2 billion for the acquisition of Reliance Electric $1 billion for development of shale oil, $0.9 billion for coal and other minerals, and $0.7 billion for uranium and uranium fabricating (*Businessweek*, 1982).

All of the investments outside of the oil and coal businesses eventually failed. Garvin terminated the oil shale project in 1982 when it became apparent that the technology was too costly to produce oil at market rates, and that government price guarantees would not be forthcoming. Moreover, after encountering perennial losses at Reliance Electric and in information processing, Garvin decided to sell these two businesses. Further, after losing $706 million on minerals and $211 million on uranium, Garvin shut down those ventures. Finally, large investments in offshore drilling in the Baltimore Canyon and Gulf of Mexico proved to be dry holes, while industry sources suggested that said Exxon had overbid on offshore properties by $700 million (*Businessweek*, 1982).

Exxon encountered many management problems in their diversification effort. First, they overestimated the future price of oil and underestimated environmental concerns, government commitment, and technical requirements of producing oil from shale. Second, they tried to manage information processing as they would an oil investment, did not scrutinize spending at the information processing unit, and did not integrate the disparate business lines into a unified structure (Harris, 1984). Third, the uranium and copper mines proved too costly and the markets too volatile for Exxon management to efficiently manage (Cook, 1985).

Recognizing shareholder discontent throughout the oil industry and the costliness of the diversification effort, Garvin terminated the conglomerate growth strategy in 1983 and began to remove Exxon from all nonoil businesses except coal. Moreover,

---

[2]Chairmen Edward Jefferson of DuPont regarded Continental as a way to insulate his company from downstream swings in chemical prices that stemmed from changes in crude oil prices (*Fortune*, 1981b).

from 1983 to 1988, Garvin and his successor, Larry Rawl, repurchased $9.6 billion in stock and repositioned the company in oil, replacing more oil than it consumed in four of the five years. Finally, by 1988, Rawl had refocused Exxon on oil, chemicals, and coal production, returning it to the businesses that it had operated in 1973.

8.1.8 *Getty Oil.* Because of its domestic oil reserves, Getty Oil was well positioned to benefit from the oil price hikes of the 1970's. After the first price hike in 1973, Chairman J. Paul Getty gained complete control of Mission Oil by acquiring the remaining stock that Getty Oil did not own. Afterwards, Getty became a completely unified company since 1937 and still had a level of debt.

A change to a diversified growth strategy in 1980 was motivated by two events. First, founder J. Paul Getty, died in 1976. Second, the combination of the oil price increase in 1979 and oil price decontrol enabled Getty to obtain $21 per barrel of oil in 1980 versus $5.40 the year earlier (*Businessweek*, 1980b). As a result, cash flow at Getty doubled while debt was low. Chairman Peterson used his strong financial position to purchase ERC Corporation, an insurer, for $587 million, and Reserve Oil and Gas for $631 million and began searching for a forest products company to acquire.

Gordon Getty, the major shareholder of Getty Oil, did not support the new diversification strategy. Upset over the drop in the stock price from $108 per share in 1980 to $65 per share in 1983 and the replacement of only 33 percent of the reserves that Getty Oil had consumed between 1978 and 1983, he argued that Peterson allocated too little cash flow for dividends and reinvestment in oil production and too much money for diversification (*Businessweek*, 1983b). Moreover, *Businessweek* reported that Gordon Getty rejected the royalty trust concept as a way to maximize shareholder value. Rather, Getty teamed up with J. Hugh Liedtke of Pennzoil to bid $110 per share for the stock that Gordon Getty did not own. After the board rejected the offer, Getty and Liedtke raised their bid to $112.50 but lost the takeover effort to Texaco, which bid $125 per share.

8.1.9 *Gulf.* After it lost its Kuwaiti and Venezuelan oil concessions in 1975, Gulf recorded one of the worst oil reserve falloffs in the oil business and yet, because of the high prices of oil, generated huge amounts of cash flow from its domestic oil resources. In responding to the new position of Gulf and the new outlook for oil, Chairman Dorsey planned to consolidate his oil business and diversify into new businesses. First, he eliminated 20 percent of Gulf's service stations, wrote off other assets, and reduced debt to 14 percent of capital. Second, he planned to exploit both the borrowing power and the high levels of cash flow of Gulf to diversify outside of oil. Dorsey, however, proved to be a poor diversifier, failing to take over CNA Financial and Rockwell International.

A political payments scandal led to the departure of Dorsey and a change in strategy. The new chairman, Jerry McAffee, discarded the diversification plan and refocused Gulf on oil and gas, chemicals, coal, and uranium. In accordance with this plan, he bought Kewanee Industries, a specialty chemicals and oil producer, while devoting most of capital expenditures to oil exploration.

Chairman James Lee, who took control of the company in 1981, accelerated the repositioning of Gulf undertaken by McAffee. He sold the European marketing and refining operation, cut employment by 25 percent, dropped low profit chemicals, and, seeking a big oil field, increased exploration in high-risk arctic and offshore ventures. Although, these ventures appeared to make sense from a long-run perspective, investors valued Gulf at one-half of its break-up value (*Businessweek,* 1983a).

After losing in a bid for Cities Service, but realizing a significant capital gain from his investment, Boone Pickens began buying stock in Gulf in 1983. If he won, Pickens planned to convert Gulf into an operating company and a Royalty Trust.5 Chairman Lee of Gulf, however, felt that Royalty Trusts were unproductive and won an initial victory over Pickens by reincorporating Gulf in Delaware, a state that requires the consent of a majority of stockholders to approve a new board of directors (*Businessweek,* 1983a).[3]

The more favorable Delaware law, however, did not prevent a takeover of Gulf. After Lee turned down an offer of $70 per share from ARCO when Gulf shares were trading at about $45 per share, Pickens, feeling that he could win a shareholder vote, bid $65 per share for the company. Recognizing that he would lose control of the company, Lee solicited bids for Gulf from ARCO, Chevron, Standard of Ohio, and Allied. Chevron won the bidding contest with an offer of $80 per share or $13.2 billion for the entire company.

8.1.10 *Kerr-McGee.* Because Kerr-McGee had interests in many businesses besides oil and produced only about 19 percent of its oil from domestic oil wells, it benefited less than other oil companies from the price increase in oil that began in 1973. Moreover, because of its existing diversified activities, it chose to concentrate its investments on existing businesses rather than expansion into new businesses and was never seriously challenged in a takeover attempt. Like many of its larger competitors, however, managers bought back stock in the mid-1980's, exited several businesses in the late 1980's, and refocused the company on oil production by 1990.

8.1.11 *Marathon Oil.* Having 47 percent of its refinery needs met by domestic production and the remainder from foreign sources, the 1973 price increase greatly enhanced cash flow at Marathon. Unlike many of their competitors, managers at

---

[3]Royalty Trusts channel money from oil producing wells directly to shareholders, bypassing both management and corporate taxes.

Marathon focused mainly on oil exploration, concentrating expenditures on oil production developments in the North Sea and the Gulf of Mexico and the acquisitions of Pan Ocean Oil and ECOL Ltd. They did, however, make small internal investments into minerals production with projects in coal, uranium, and other minerals.

Managers used the proceeds stemming from the second major oil price hike of 1979 for capital expenditures, allocating $960 million in 1981 versus $500 million in 1978 and $350 million in 1976. Much of the new investment went for development projects in the North Sea and Gulf of Mexico. Investors, however, did not agree with the use of cash flow, placing a $62 per share value on the stock relative to what some analysts believed was a $125 per share break-up value (*Businessweek*, 1981).

Managers at Mobil Oil, searching for sources of oil located in safe political climates, reacted to the low Marathon stock price by making an offer of $85 per share in late 1981. Fearing a loss of their control, Marathon managers, allied with independent dealers, won a court injunction against further actions by Mobil and solicited offers from other companies. Marathon eventually reached an agreement with U.S. Steel for $6 billion, about $900 million more than Mobil.

8.1.12 *Mobil Corporation.* Fearing political unrest in the countries in which it obtained oil and wanting to limit its exposure to possible nationalizations abroad, Mobil established a policy goal of diversifying into new businesses in 1968. After the oil price hikes of 1973 and the beginning of oil property nationalizations by many oil producing countries, management at Mobil intensified their search for a suitable hedge against risk exposure abroad. In 1976, Chairman John Warner purchased Marcor, a company with businesses in retailing through Montgomery Ward and containers with Container Corporation.

The acquisition of Marcor may have seemed to be a good one, coming at a cost $1.8 billion for a company that generated $150 million in profits per year. Throughout the 1970's, however, profitability at the Montgomery Ward unit declined until in 1979 it began losing money. Mobil managers, lacking expertise in running a retailer, could not find a manager who could turn the retailing unit around and lost $512 million, making $609 million in investments in the division (Siler, 1987). Finally, Mobil managers appointed Bernard Brennen as president of the retailer in 1985, who returned Montgomery Ward to profitability after he closed the unprofitable Jefferson Ward discount chain, shut down the catalogue business, and eliminated several product lines.

After the acquisition of Marcor, Warner concentrated on obtaining more crude oil. In the late 1970's, he bought General Crude and Trans-Ocean, two crude oil producers, for $1.5 billion, increased Mobil's share of ARAMCO output to 15 percent, and increased exploration spending by 50 percent. The exploration strategy of focusing on only large offshore fields, however, failed to keep pace with other oil company exploration efforts and yielded the lowest number of exploratory wells among the major oil companies.

In spite of its increased spending on exploration and crude oil acquisitions, the oil reserves of Mobil declined at a rate of 10 percent per year from 1978 to 1981. In accordance with its position in crude oil and the change in political administrations to one more friendly to energy company mergers in 1980, Mobil actively pursued acquisitions of CONOCO, Marathon, and Cities Service but had no success until it acquired Superior Oil for $4.6 billion in 1984.

After its acquisition of Superior Oil, managers at Mobil undertook plans to consolidate Mobil holdings. First, they sold Container Corporation of America and, later, the now profitable Montgomery Ward unit. Second, investment was concentrated in areas where Mobil already had strong positions. Consequently, Chairman Allen Murray upgraded service stations and concentrated resources on the eastern states, while abandoning the northwest market. Moreover, he cut management ranks and continued the program begun by Warner of eliminating inefficient refineries. After making these moves, Mobil once again became only an oil and chemical company.

8.1.13 *Occidental.* Having a large proportion of its oil producing properties in Libya, Occidental lost many of its assets when Libya nationalized oil production in the early 1970's. As a consequence, Chairman Armand Hammer concentrated on oil exploration in the 1970's, while shutting down the unprofitable refining operation. Occidental recovered from their setback in Libya by focusing its resources on foreign oil exploration and made rich oil strikes in Peru and other locations.

The strategy changed in 1981. First, claiming that Iowa Beef Processors provided another service to farmers and thus complementing the fertilizer and pesticide businesses of Occidental, Hammer acquired the meat-packing company in 1981. Next, seeing an opportunity to give Occidental a domestic oil production base and large tracts of domestic oil properties to balance international operations, Hammer acquired Cities Service in 1982 for $4 billion and thus, for the first time since the 1960's, primarily became a domestic oil company. To partially pay for the acquisition, Hammer sold the refining and marketing system of Cities Service. Another acquisition came in 1986, when Hammer, observing that gas production from Cities Service lacked a secure market, acquired Mid-Con, a natural gas distributor, for $3 billion. Not yet satisfied with the size of the company, Hammer bought Cain Chemical in 1988 for $2.2 billion in an effort to obtain its ethylene production for the chemical unit.

Occidental differed from the other large oil companies during the 1980's in that shareholders exerted little pressure on management to enhance value. Yet, the stock traded at only 40-50 percent of break-up value in 1988 (Kirkpatrick, 1988). Several factors may have deterred takeover specialists. First, Hammer maintained a debt load of 50-60 percent of capital. Second, Hammer instituted a dividend with a 10 percent yield. Third, Hammer had a great deal of influence over the board of directors; hence, a takeover might have been costly. Fourth, until Hammer purchased Cities Service, Occidental had little domestic oil production; thus, since both raiders and managers at other domestic oil companies sought domestic oil reserves, Occidental was not an attractive takeover target.

8.1.14 *Phillips Petroleum.* After adopting the multidivisional form in 1975, managers at Phillips expanded coal and uranium operations and continued domestic exploration. Unlike managers at other oil companies, however, they did not pursue a broad diversification program. Rather, they chose to reduce debt, resulting in a level of 13 percent of capital by 1981. Maintaining their strategy of focusing on the oil industry but recognizing the need for both domestic sources of oil and greater debt to make the company less attractive as a takeover target, managers at Phillips bought General American Oil and Aminoil in 1983 for $2.9 billion.

The effort to prevent a takeover threat did not succeed. With its stock trading at $38 and a break-up value believed to be about $80 per share, Boone Pickens began acquiring Phillips stock and made a $60 per share offer in 1984 (Norman, 1985). Pickens, however, relented when Phillips undertook a recapitalization that analysts valued at $53 per share and cost $4.3 billion. The resulting debt of 80 percent of capital compelled Phillips management to sell its shale oil, coal, and uranium assets, reduce employment from 29,000 to 20,000, and cut capital expenditures on oil by 50 percent. Moreover, it was not until Phillips realized strong performance in its chemical operations unit at the end of the 1980's that Phillips management was able to eliminate the possibility of not meeting debt payments and could concentrate on developing the chemical and oil businesses.

8.1.15 *Shell.* Shell devoted most of its efforts to its traditional oil and chemical businesses during 1970-84. Its only diversification effort came between 1968 and 1973 when it became a partner with Gulf in a joint effort to develop breeder reactors. Encountering little success, however, both Shell and Gulf abandoned the project in the early 1970's. Afterward, Chairman John Bookout concentrated on growth in oil and chemicals.

Shell proved to be a particularly good production company during the 1970's, realizing the highest crude replacement rate and lowest finding cost of any of the major oil companies between 1978 and 1982. Aside from efforts to increase oil production, Shell also spent $2.5 billion between 1975 and 1979 to expand its chemical capacity into olefins, aromatics, and proprietary chemicals.

The only major acquisition by Shell came in 1979 when Bookout, believing that Shell could obtain oil for $3.50 per barrel, paid $3.65 billion for Belridge Oil (*Forbes*, 1980). The combination matched the technical skills of Shell with the heavy crude oil of Belridge to make a good fit and later proved successful, as Shell doubled output from the wells by 1984.

Although Shell had much success through its independence from Royal Dutch Shell, the British-Dutch company acquired the 30 percent of Shell's stock that it did not own in 1984 as way to better control investment spending, use of cash flow, and reduce management redundancies *(Businessweek,* 1984b). Moreover, strategic-analysis chief E.V. Newland suggested that Royal Dutch Shell would have made the acquisition much earlier but did not do so because of fear of anti-trust actions *(Businessweek,* 1984b).

8.1.16 *Standard Oil of Ohio.* After his exchange of control of the company for production acreage in Alaska with British Petroleum in 1969, Chairman Charles Spahr devoted most of the resources of Standard of Ohio to its development of the oil properties. Having insufficient cash flow to develop the oil properties without outside funding, Spahr financed the investment with debt, which reached a level of 74 percent of capital by 1976. The vast oil resources of Standard of Ohio and the combination of oil price hikes and decontrol of oil prices in 1979, allowed Standard of Ohio to sell 600,000 barrels of oil per day at a gross margin of $16.50 per barrel in 1980.

In spite of both a weak exploration department and poor performance in past diversifications into coal, chemicals, and hotels, Chairman Alton Whitehouse chose to emphasize exploration and to obtain 25 percent of revenues from nonoil businesses *(Businessweek,* 1980c). Initially, Whitehouse used the money to reduce debt to 30 percent of capital, increase the size of the Standard exploration department, purchase domestic oil properties, and improve prospects in the chemical and coal businesses. In spite of these spending initiatives, Standard still managed to accumulate $2.5 billion in marketable securities by 1980, up from $0.2 billion in 1978. In response to the windfall of available spending power and after pressure from Congress to avoid nonenergy acquisitions had subsided, Chairman Whitehouse chose to diversify into either nonenergy minerals, chemicals, genetic engineering, or semiconductors. He settled on a $1.7 billion acquisition of Kennecott, a mining company, and coal properties valued at $805 million.

The exploration and diversification program undertaken by Standard of Ohio met with little success. From 1978 to 1984, Standard had the second highest domestic average finding cost of new oil and replaced only 16 percent of the oil it had produced. Moreover, between 1980 and 1984, Kennecott lost $537 million. Feeling no pressure from British Petroleum to change, Whitehouse maintained his business strategy in spite of a stock value worth 60 percent of the value of the properties in 1985 (Simon, 1985).

British Petroleum finally exerted pressure on Standard in 1986, appointing Robert Horton from its own operations to replace Whitehouse. His immediate plan was to cut exploration expenditures and refocus Standard on its refining and marketing strengths. In 1987, seeking complete control of Standard and wanting to consolidate its U.S. holdings, British Petroleum bought the remaining stock that it did not own.

8.1.17 *Sun Oil Company.* Like many of its competitors, Sun benefited from the oil price increase in the early 1970's. Being short on crude oil, however, Sun realized far fewer gains. With the cash that it did have, first the Pew family and, later, Chairman H. Robert Sharbaugh decided to emphasize production of oil from tar sands in Canada. Having little success with the technology, Chairman Sharbough reorganized the company into six operating units, each able to act independently of the other, to serve as a vehicle through which he planned to diversify Sun. He felt that structure would enable him to be able to quickly evaluate profitability of individual units, divest parts of the company that were not profitable, and facilitate the management of nonoil companies (*Businessweek*, 1976). Speaking of his new business strategy, Chairman Sharbough stated, "The time to start diversifying is now while we enjoy good cash flow and can afford to take risks."

In the subsequent diversification program, Sharbough took Sun into trucking, industrial distribution, computers, electrical equipment, coal, mining, and other more minor ventures while simultaneously disengaging Sun from oil exploration. One of the ventures, a 34 percent interest in Becton Dickenson, proved costly to both Sun and Sharbough. After making the purchase, the SEC charged that Sharbough did not make the 37 percent premium, which he paid to private investors, available to all shareholders. After a settlement in which Sun divested itself of Becton stock, Sharbough lost his job.

Sharbough's strategy had resulted in a crude self-sufficiency ratio of 33 by 1977, dropping from almost complete self reliance in 1965. New chairman, Theodore Burtis, reacted to the deteriorating crude position by buying Texas Pacific for $2.3 billion in 1980 and thereby raising reserves by 23 percent. In an effort to consolidate operations and refocus Sun in the oil business, Burtis pulled Sun out of 26 of the 36 states in which it marketed oil, exited many of the businesses that Sharbough had entered, cut the labor force by 10.5 percent, and sold 1,400 marginal production properties. In spite of its renewed emphasis on oil production, Sun replaced only 59 percent of its oil between 1982 and 1989, while incurring a finding cost of $8.07 per barrel versus the $6.89 industry average cost (Mack, 1988).

Burtis also made two efforts to better reward Sun shareholders. First, he spent $645 million for the repurchase of Sun stock in 1985. Second, feeling that investors would value Sun greater as two companies, he split Sun into an exploration company and a refining firm and spun both off to shareholders in 1988.

8.1.18 *Tenneco.* The strategy of Tenneco changed little from 1970 to 1986. It remained a highly diversified company but did face criticisms among oil company analysts in 1981. These analysts believed that Tenneco should spin off the oil business in order to generate more value for the oil properties, which exceeded those held by either Amoco or Texaco (*Businessweek,* 1981b). Rather than reorganizing, however, Chairman James Ketelson imposed an austerity program on the auto parts, farm machinery, packaging, and chemical sub-units, while using 60 percent of cash flow for oil exploration and production, international exploration, and synthetic fuels. Investors agreed with the analysts view of reorganizations, valuing Tenneco at one-half the value of its assets. Nonetheless, Ketelson placated shareholders with a 33 percent of cash flow dividend and thus did not encounter takeover threats (*Businessweek,* 1981b).

The investments made in the early 1980's proved costly. The synthetic fuel business was eventually abandoned while, speaking for other projects, Executive Vice-President Joe Foster admitted to overbuying oil properties (*Businessweek,* 1984d and 1985a). Moreover, the acquisition of the farm machinery and construction equipment business of International Harvester yielded little return as the farm machinery business failed to turn around.

After losing $259 million in 1986-87, having a debt load of 66 percent of capital, and recognizing that Tenneco stock traded at one-half of asset value, Ketelson decided to exit the insurance and oil businesses and refocus Tenneco on manufactured products and natural gas distribution.

8.1.19 *Texaco.* In spite of owning significant quantities of domestic oil properties, Texaco did not realize the financial gains that many of the other major oil companies realized. Oil property nationalizations by other countries, the decline in Texaco's domestic oil reserves, previous investments in small inflexible refineries, and government price controls combined to displace Texaco from industry leadership in terms of return on equity in 1967 to last place in 1977.

Many of the problems Texaco encountered stemmed from the administration of Gus Long, who retired as Chief Executive Officer in 1972. He based company strategy for 20 years on the pre-1973 position of Texaco and the world petroleum market in which transportation costs were relatively high and the company had access to plentiful supplies of domestically owned oil of consistent quality. As the oil reserves of Texaco became depleted and Saudi Arabia asserted itself in the ARAMCO partnership, however, Texaco was forced to buy high-cost special grades of oil for its inflexible refineries. Investments in small refineries proved to be costly as transportation costs dropped, enabling a larger refinery to serve a broader market. Further, government price ceilings reduced the price that Texaco could obtain for already discovered domestic oil, thereby limiting the cost advantage Texaco had over nonoil producing refiners. Finally, management had not participated in Alaskan, North Sea, and other major production fields, resulting in high-cost current operations with bleak future prospects.

In response to these circumstances Maurice Granville, Long's successor, undertook a strategy of eliminating marketing operations in several western states, closing the smallest refineries, and upgrading other refineries, while increasing expenditures on oil production. Renewed emphasis on oil production was not successful, as Texaco reported one of the poorest crude replacement rates in the industry. In an effort to gain more crude and perhaps as a way to increase debt and limit its own exposure to a takeover, Texaco bought Getty Oil in 1984 for $10 billion.[5] The new reserves came at a price of $4.63 per barrel and, because of it was heavy crude oil, a $10 extraction cost, versus the $19 per barrel Texaco was paying for its oil exploration efforts (*Businessweek*, 1984a). Most of the oil, however, could be sold for only $20 per barrel.

The acquisition of Getty proved to come at a much higher cost than the $10 billion originally paid. In a lawsuit over the deal, Chairman Hugh Liedtke of Pennzoil, who with Gordon Getty had made a competing offer for Getty Oil, claimed that the board accepted the offer made by Texaco after it had already agreed to the offer by Liedtke. In a lawsuit against Texaco, Liedtke won a $10.3 billion award for Pennzoil, which Texaco appealed. Further complicating Texaco's problems was a 1987 acquisition of 12.3 percent of the company by Carl Icahn, a takeover specialist.

After pressure from investors and Icahn, James Kinnear, the new chairman of Texaco, settled with Liedtke for $3 billion in 1988. Icahn, however, was not satisfied. With the stock value of his holdings trading at $51 per share but with an asset value of $67 per share, Icahn attempted a takeover (Sherman, 1989). He lost, but only after institutional investors had asserted their power by forcing a change in company by-laws that made takeovers easier and extracting a one-time dividend of $1.7 billion.

The dividend payment, the settlement with Liedtke, and the heavy debt load, which the company had carried since its acquisition of Getty, forced Kinnear to sell assets. He sold Texaco-Canada to Exxon for $3.2 billion, Texaco's West German subsidiary for $1.2 billion, one-half interest in the East Coast refining and distribution operations, and other assets. Moreover, Kinnear streamlined the Texaco organizational structure by eliminating two layers of management and reducing the payroll by 30 percent. Finally, a revised oil production strategy yielded one of the lowest finding costs of oil in the industry at $3.33 per barrel.

8.1.20 *UNOCAL*. Having ample reserves of domestic crude oil supplies, UNOCAL was well positioned to take advantage of the price increases that occurred during the 1970's. The availability of cash flow gave Chairman Fred Hartley the options to reduce debt, diversify, or increase dividends. He chose to keep dividends at about 25

---

[5]Businessweek asserted that Texaco may have paid the high price it did as a way to avoid a takeover.

percent of cash flow, reduce debt, and purchase Molycorp, a mining company, for $280 million.

The strategy backfired in 1985 when Boone Pickens accumulated 13 percent of UNOCAL stock and threatened a takeover. Norman and Starr (1985) reported that Pickens planned to offer $55 per share, a $23 increase from the traded price that existed when he began accumulating stock. In order to avoid a takeover, Hartley increased debt from 18 percent to 75 percent of capital by repurchasing one third of Union stock for $4.2 billion.

The heavy debt load that UNOCAL incurred constrained Hartley to grow internally rather than through the acquisition of cheap oil properties that came on the market in late 1980's. Facing a dropping crude price and a termination of government price guarantees on synthetic oil, Hartley reduced the value of the synthetic fuel business. Richard Stegemeir, who replaced Hartley in 1988, accelerated the pace of debt reduction by putting geothermal and real estate assets on the market, while shutting down the shale oil project and refocusing the refining and distribution operations. As a result, UNOCAL repositioned itself as a largely west coast refiner and marketer with ample supplies of oil.

8.2 Discussion and Conclusion

Oil company managers pursued either related or unrelated growth strategies in the period following 1973. Two reasons appear to explain the decision to diversify. First, managers sought a way to reduce reliance on earnings based on overseas operations. For example, managers at Sun and Ashland were particularly fearful of losses because of their poor position in domestic oil, while management at Mobil sought to reduce their reliance on earnings from overseas ventures.

Second, managers sought investment outlets for use of their high cash flow but, lacking opportunities in domestic oil production, used cash for alternative investments. Standard of Ohio, for example, had $2.5 billion in marketable securities in 1980 before it bought Kennecott for $1.7 billion. Moreover, of the six major oil company acquisitions listed in table 8.1, four immediately followed the second major price hike in 1979 and two came after the first increase in 1973.

Managers choosing only related growth appear to have done so as way to either correct management mistakes made in the past or to finance ongoing projects. Gulf and Texaco, for example, had previously invested in an inflexible refining and distribution network, that relied on high quality grades of oil, which were both in short supply and more costly. Hence, these companies had to upgrade their operations or close them down. Cities Service, CONOCO, and Kerr-McGee, on the other hand, used cash flow to further develop businesses into which they had previously invested (such as copper, coal, and minerals).

Oil company managers did not successfully diversify into new businesses. By 1990, all companies except Occidental had returned to almost the same lines of businesses in which they had operated in 1970 (table 8.2). The problems that managers encountered in their diversification drive were numerous. First, managers did not act as if they were discriminating investors, appearing instead to view new ventures as a way spend surplus cash. Exxon, for example, provided such an excess supply of investment funding to their information processing unit that managers at the subsidiary attempted more projects than they could manage efficiently, eventually failing in all of them (Ticer and Glasgall, 1985). Second, an incompatibility in bureaucratic response time also proved costly. Because Exxon centralized control over investment projects and the Exxon system was designed to accommodate projects that required seven years to return a profit, managers could not act quickly enough to respond to a dynamic market (Harris, 1984).

Even under conditions of comparable bureaucratic processes and a match of a company with surplus cash and a recipient with insufficient cash, managers were not able to profitably develop their diversified businesses. For example, managers at ARCO and Standard of Ohio bought mining companies with compatible financing needs and similar bureaucratic structures, but they were unable to recognize market trends toward lower-cost foreign copper sources and the high-cost nature of the purchased mines. Other cases of a lack of management expertise included Mobil, which lost money in retailing and containers, Ashland, which lost money in pollution control equipment and insurance, and Exxon, which incurred losses approaching $3 billion in information processing and mining.

Unfamiliarity with the operations of new businesses, also led oil company managers to make poor choices of subordinates at their acquisitions. Mobil, for example, took seven years before they could find a manager who could turn around their Montgomery Ward retailing unit.

Although they did not incur the costs of diversification that other companies experienced, managers that maintained a related growth strategy, like managers that followed an unrelated strategy, failed to provide value to shareholders. CONOCO, Gulf, and Marathon, for example, were acquired by other companies mainly because their stock values traded at a low price relative to their asset value. Shareholders at Texaco, on the other hand, extracted a $1.7 billion dividend only after the company had been threatened with a takeover. All companies except Mobil, Occidental, Chevron, Standard of Ohio, and Shell made stock value enhancements between 1970 and 1990, and most came only after shareholder pressure (table 8.3). Mobil, Occidental, and Standard may have avoided the payment to shareholders because of their large debt loads, while Shell and Standard of Oil were controlled by other companies.

Anti-trust enforcement may have played a role in the industry consolidation of the 1980's. All of the major horizontal acquisitions came in the 1980's after a change of federal political administrations in 1980 (table 8.2).

These observations suggest two attributes of the multidivisional structure. First, because they lacked knowledge about the industries into which they entered, top managers relied on subordinates to operate the diversified businesses. Yet, this reliance on subordinates eliminated any benefit from central office monitoring and interjected a bureaucratic structure (the central office) between a discriminating capital market (the external market) and the operating unit.

Second, managers derived the full benefit from the multidivisional form through highly related investments in which natural complements existed with the previous business lines. In the absence of the benefits achieved through centralized control, however, the multidivisional form loses its chief advantages and becomes an internal mechanism for allocating capital, which, for oil companies, proved to be inefficient.

TABLE 8.1–ACQUISITIONS BY MAJOR OIL COMPANIES 1970-90

Holding Company	Year	Acquisition or Sale	Value[1]	Type	Industry
ARCO	1977	Anaconda	500	U	Mining
Cities Service	1982	-	4,050	H	Occidental buys
CONOCO	1981	-	-	H	DuPont acquires
Exxon	1979	Reliance	1,236	U	Electric motors
	1987	Celeron Oil	650	H	Oil production
	1989	Reliance	(1,700)	U	Electric motors
	1989	Texaco-Canada	4,100	H	Oil production
Getty	1980	ERC	586	U	Insurance
	1981	Reserve	628	H	Oil production
	1984	-	(10,000)	H	Texaco acquires
Gulf	1984	-	(13,200)	H	Chevron buys
Kerr-McGee	-	-	-	-	-
Marathon	1981	-	-	H	U.S. Steel buys
Mobil	1974	Marcor	1,600	U	Retail/containers
	1979	General Crude	792	H	Oil production
	1980	Trans-Ocean Oil	715	H	Oil production
	1982	Ansutz	500	H	Oil production
	1984	Superior Oil	5,720	H	Oil production
	1986	Montgomery Ward	(1,600)	U	Retailing
	1988	Container Corp.	(700)	U	Containers
Occidental	1981	Iowa Beef	771	U	Beef processing
	1982	Cities Service	4,050	R	Oil/natural gas
	1986	Mid-Con	3,000	R	Natural gas
	1986	Dia. Shamrock Chem	850	R	Chemicals
	1988	Cain Chemical	2,300	R	Chemicals
Phillips	1983	General American	1,140	H	Oil production
	1984	Aminoil	1,600	H	Oil production
Shell	1979	Belridge	3,653	H	Oil production
	1985	Royal Shell buys	-		-
Chevron	1984	Gulf Oil	13,300	H	Oil production
	1988	Tenneco properties	2,600	H	Oil production
Amoco	1988	Dome Oil	4,200	H	Oil production
	1988	Tenneco properties	900	H	Oil production
Standard of	1981	Kennecott	1,770	U	Mining
Ohio	1981	U.S. Steel Coal	600	R	Coal
	1985	British Petr. buys	-		-
Sun Oil Co.	1980	Seagram properties	2,300	H	Oil production
	1988	Atlantic Petroleum	594	H	Oil production
Texaco	1984	Chevron of Europe	800	H	Oil refining
	1984	Getty	10,200		Oil production
	1989	Texaco Canada	(4,100)		Oil production
UNOCAL	N.A.	N.A.	N.A.	N.A.	N.A.

[1] Figures in parenthesis represent a sale of assets; minimum value included in table = $500 million.
H = horizontal acquisition; R = Related acquisition; Unrelated = Unrelated acquisition; N.A. = not applicable.
Sources: *Wall Street Journal* 1970-90; *Moody's Industrial Manual* 1970-90.

TABLE 8.2–PRODUCT LINES OF OIL FIRMS IN TIME SERIES, 1950-90

Firm	Year				
	1950	1960	1970	1980	1990
Ashland	reg	reg	reg,io,pe,cn	reg,p,cn,sh,cl	reg,pe,cl,cn
ARCO	reg	reg	nat,pe,ch,oth	nat,io,pe,cl,m,u,s	reg,io,pe,cl,u
CONOCO	reg	nat,pe	nat,io,pe,cl,u	nat,io,pe,cl,u	bought by DuPont
Getty	reg,oth	reg,io,u	nat,io,u	nat,io,u,cl	bought by Texaco
Gulf	reg	io,pe	nat,io,pe,cl,oth	nat,io,pe,cl,u	bought by Chevron
Kerr-McGee	reg	reg,u	reg,u,m,ch,oth	reg,u,gas,m,ch,cl,oth	reg,u,ch,m,cl
Marathon	reg	reg	nat,io	nat,io,cl,u	bought by U.S. Steel
Occidental	reg	reg	reg,io,m,ch,r,cl	reg,io,m,ch,r,pe,cl,oth	nat,io,m,ch,r,pe,oth,m,cl
Sun Oil Co.	reg,sh	reg,sh	nat,sh	nat,io,sh,s,pe,oth	reg,io,cl,u,r
Tenneco	gas	gas,reg	gas,reg,sh,pe,pk,ins,ag,mn	gas,oil,sh,pe,pk,mn,ins	gas,reg,mn
Texaco	nat,io,p	nat,io,pe	nat,io,pe,ch	nat,io,pe,ch	nat,io,pe
Union	reg	reg,pe	nat,pe	nat,io,pe,m,ch	nat,io,pe,mn
Cities Service	reg	reg,pe	nat,pe,oth,m,r,a	nat,pe,cl	bought by Occidental
Exxon	nat,io,pe	nat,io,pe	nat,io,pe,cl,u,ch	nat,io,pe,cl,u,ch,m,oth	nat,io,m,pe
Mobil	nat,io	nat,io,pe	nat,io,m,pe,ch,oth	nat,io,pe,u,cl,m,oth,cn,oth	nat,io,pe,u,cl
Phillips	nat,io,pe	nat,io,pe	nat,u,cl,gas,oth	nat,u,cl,oth	nat,u,cl
Shell	nat,io,pe	nat,io,pe	nat,io,pe	nat,io,pe,cl	nat,io,pe,cl
Chevron	nat,io,pe	nat,io,pe	nat,io,pe,oth	nat,io,pe,m,cl,u	nat,io,pe,cl,u
Amoco	nat,pe	nat,io,pe	nat,io,pe	nat,io,pe,m,cl,u	nat,io,pe,cl,u
Stand. of Ohio	reg	reg,pe	nat,pe,oth,cl	nat,pe,io,cl	bought by British Petroleum

Definitions: ch = agrichemicals, fertilizers, and inorganic chemicals; cl = coal; cn = construction; gas = gas; ins = insurance; io = international oil; m=minerals; mn = manufacturing; nat = national oil; oth = nuclear fuel, iims, plastic, forestry, electronics, beef, retailing, trucking, agriculture, and aircraft; pe = petrochemicals; pk=packaging; r = real estate; reg = regional oil; sh = ships; and u = uranium.

TABLE 8.3–STOCK VALUE ENHANCEMENTS BY MAJOR OIL COMPANIES
BETWEEN 1970 AND 1990

Firm	Year	Type	Value[1]
Ashland	1979	Stock repurchase	257
	1986	Stock repurchase	
			134
ARCO	1985	Stock repurchase	5,300
		Dividend increase	150
Cities Service	1982	Sold to Occidental in 1982	4,050
CONOCO	1981	Sold to DuPont in 1981	5,800
Exxon	1985	Stock repurchase	4,300
	1982-85	Dividend increase	200
Getty	1984	Sold to Texaco in 1984	10,200
Gulf	1984	Sold to Chevron	10,600
Kerr-McGee	1985	Stock repurchase	75
Marathon	1981	Sold to U.S. Steel in 1981	5,000
Mobil	-	-	-
Occidental	-	-	-
Phillips	1985	Stock repurchase	4,300
Shell	1985	Royal Dutch Shell buys 39 percent it does not own.	5,700
Chevron	-	-	-
Amoco	1985	Stock repurchase	1,750
Standard of Ohio	1985	British Petroleum buys stock it does not own	-
Sun	1985	Stock repurchase	1,145
Texaco	-	-	1,700
UNOCAL	1985	Stock repurchase	3,600

* Values are in millions of dollars

Source: *Wall Street Journal Index* 1970-90; *Moody's Industrial Manual* 1970-90.

Chapter 9

Discussion and Conclusions

9.1 Discussion

The findings of this book suggest that single-product functionally organized firms and multidivisional multiproduct firms are both efficiently organized while decentralized firms are inefficiently structured.

9.1.1 Profitability. Theorists (Williamson, 1975; Armour and Teece, 1978) have suggested that the multidivisional form is a superior organizational structure to the functional arrangement for large firms. They also claim that the lag time between the introduction by the first user and the adoption by the last user is a result of simple diffusion. The results presented here, however, suggest otherwise. Empirical tests of long-run profitability (tables 7.5, 7.6, 7.7, 7.9, 7.10, 7.11, and 7.12), operating profitability (table 7.13), and managerial efficiency (table 7.14) suggest that the multidivisional form is not unambiguously superior to the functional form. Rather, it is a more efficient way to organize only for multiproduct firms.

Table 7.5 indicates that multidivisional firms were more profitable than functional firms from 1954 to 1961 and less profitable in other periods. Decentralized firms, in contrast, were significantly less profitable than functionally organized firms throughout most of the 1950 to 1972 period.

These results can be explained as follows. Table 7.8 indicates that the early multidivisional firms were among the first firms in the oil industry to enter international markets, petrochemicals, uranium, or all three. As a consequence, these firms were able to acquire concessions for producing oil on the most lucrative tracts of land and able to earn economic rents in excess of $1.50 per barrel over domestically produced oil during the 1950's (Burrows and Domencich, 1970). Moreover, the first multidivisional firms were able to exploit their research and development in petroleum products into the production of petrochemicals, when that industry was in its early growth period and offered high margins.[1] Further, some oil companies were able to take advantage of exploration expertise through entry into uranium when the Atomic Energy Commission was offering contractual guarantees of profitability.[2] Finally, most of the early multidivisional firms were fully integrated by the 1950's, thereby reducing their reliance on uncertain long-term contracting. Hence, it was likely that the moderate profits for multidivisional firms during the earlier period of the study was a result of market activities.[3]

During the 1960's, many firms began entering both the petrochemical and international oil businesses (table 7.4). As a result, oil price cost margins dropped in foreign markets to about $0.35 per barrel and petrochemical profits deteriorated (Burrows and Domencich, 1970). Moreover, because of the Oil Import Quota, firms with low-cost sources of domestic oil became more profitable. Hence, results for single-product functional firms and multiproduct multidivisional firms converged.

The significant drop in profitability for multidivisional firms during the late 1960's and early 1970's can be explained by three factors. First, Libya and, later, the other oil producing countries doubled the tax they imposed on oil producers in 1969. Second, several companies began diversifying into unrelated businesses in which firms did not have a significant competitive advantage (Montgomery and Wernerfelt, 1988) Third, Schipper and Thompson (1983) believe that the Williams Act of 1969 raised the cost of diversification.

Product-related interpretations cannot be made for decentralized firms. If market activities were a source of economic rents in the earliest period, then decentralized firms should have outperformed functional firms. Moreover, since in other periods their market activities are similar to functionally organized firms, their performance should not have been different. Yet, they underperformed functional firms in later periods and multidivisional firms in the earliest period.

The finding that Tobin's Q declined for six functionally organized firms during their study periods may have been driven by the mode of diversification (Table 7.4).

---

[1]Prices dropped by about 66 percent for a typical barrel of petrochemicals between 1950 and 1970.

[2]Owen (1985) indicates that the AEC made payments of $8.00 to $12.51 per pound of uranium from 1950 to 1962, but then reduced payments to $8.00 thereafter. He also indicates that the average cost of production during this time was about $5.00 per pound.

[3]Montgomery and Wernerfelt (1988) found that sales in international markets temd to increase Tobin's Q. Moreover, diversification into petrochemicals offered competitive advantages to oil refiners because effluent gases could be used to produce a valuable good.

According to Montgomery and Wernerfelt (1988), because firm specific skills can be transferred more readily into businesses that are similar to the primary business of the firm, diversification into unrelated businesses leads to a decline in Tobin's Q while growth into international and regional markets leads to an increase in Tobin's Q. In accordance with this finding, of the six firms that had lower rates of long run profitability, Occidental, Tenneco, Sun, Ashland, and Belco diversified into non-petroleum markets, while Texaco changed organizational form just prior to World War II. Moreover, of the three firms that realized gains in profitability, both Gulf and Marathon entered international markets, while ARCO grew into a nationwide firm. Hence, product diversification appears to have driven the results.

The results of empirical tests of decentralized firms suggest that these firms had previously been inefficiently organized prior to their use of the multidivisional form. Table 7.7 indicates that decentralized firms improved their profitability after adopting the multidivisional form. Moreover, Table 7.5 shows that decentralized firms consistently underperformed functional and multidivisional companies during 1950-70.

Table 8.2 indicates the extent of diversification by oil companies. The historical evidence provided in Chapter 4 suggests that because the multidivisional form facilitates the measurability of subordinate performance and enables managers to consolidate and centralize functions common to all sub-units, diversification into businesses that have natural complements with the oil industry and fit well into the multidivisional form. The historical information provided in Chapter 8, however, suggests that because few natural linkages existed between unrelated sub-units and current lines of business and central management was unfamiliar with the markets into which they diversified, unrelated growth was not successful. As a result, most firms had come almost full circle by 1990 and had returned to the same businesses in which they were active in 1973 (table 8.3).

9.1.2 *Other Issues.* Consider several issues that both Williamson (1975) and Armour and Teece (1978) raise. First, they argue that radial expansion results in cumulative control loss and the confounding of strategic planning and operating decision-making. The data, however, indicates that large diversified firms did not grow radially. Rather, they grew laterally by either merging with other firms and organizing them as subsidiaries or creating decentralized units for new product lines. In this framework, top management still focused its efforts on the primary market. The multidivisional form acted to consolidate operations by centralizing similar activities, appointing a sub-unit manager to oversee the primary market, and finally adding a higher level of management to monitor all sub-unit managers and focus more top management time on secondary markets. Hence, in contrast to what Williamson (1975) and Armour and Teece (1978) argue, the multidivisional form grows radially upon the existing structure and can therefore increase organizational costs if improved operating efficiency from centralized control does not occur.

Second, Armour and Teece (1978) assert that the delay of the spread of the multidivisional form was a result of simple diffusion. The year of organizational change by each firm (tables 7.9 and 7.13), however, bears little relation to performance for functionally organized firms. Table 7.5 does indicate that higher profits existed for the early multidivisional firms, but it also suggests that these firms were significantly less profitable during later periods, implying that it was not organizational form that drove the results but some other consideration, such as business activities. Finally, there is a 35-year lag between Texaco, the first firm to adopt the multidivisional form, and Phillips, the last. If competitive forces were driving diffusion, and the organizational technology was superior, one would expect a very rapid diffusion.

Third, Armour and Teece (1978) also believe that size alone drives the decision to divisionalize. Exxon, however, the largest firm in the sample, did not change organizational structure until 20 years after Texaco and 40 years after General Motors. Moreover, other large firms such as Gulf, Mobil, Phillips, Amoco, and Sun waited between 16 and 33 years to adopt the new organizational technology, while much smaller firms, such as Kerr-McGee, CONOCO, and Getty, converted prior to these much larger firms, suggesting that another motive for multidivisionalization was present.

Based on the historical and empirical evidence, it appears that functionally organized oil companies used the multidivisional framework as a better way to accommodate growth in secondary markets. But tests neither indicate that the multidivisional form allows the firm to attain a higher level of efficiency nor is beneficial to investors than a single-product functional firm. Rather, it allows a single-product firm to expand its organizational scope. Contrary to Williamson (1975) and Armour and Teece (1978), the 34-year period between the first multidivisional oil firm, Texaco, and the last, Phillips, is not a simple diffusion process but a result of a firm specific decision to expand the scope of its operations. This finding is consistent with Rumelt (1986), Chandler (1977), and historical data suggesting that the multidivisional form is adopted by multiproduct firms, those seeking to diversify, and those seeking growth.

The findings of this study are consistent with other empirical findings by Hill (1985), Steer and Cable (1978) and Thompson (1981) who compared holding company performance to that of multidivisional firms. Moreover, the implication that the functional organizational structure is efficient for single-product firms may explain the ambiguous results of Cable and Dirrenheimer (1983), Teece (1981), Armour and Teece (1978), and Harris (1983).

9.2 Conclusion

The findings of this book indicate that the work of Armour and Teece (1978) was incomplete and, at best, the multidivisional form yields only transitory profit gains and no gain in long run profitability for functionally organized firms. These results are robust across capital market and accounting data measures. It may be that competitive conditions were so strong that any efficiency gains attributed to the multidivisional form quickly dissipated.

The main conclusion presented here is that single product firms are efficiently organized as functional firms. However, they may lose their efficiency after diversifying. Accordingly, they adopt the multidivisional form as a way to more efficiently manage diverse market activities. Empirically, this is supported by significant results for growth of sales in diversified markets and efficiency improvements only for decentralized firms.

Appendix

The Oil Industry from 1950 to 1990

A.1 Overview

The years from 1950 to 1990 can be decomposed into two distinct periods, one lasting from 1950 to 1973 and the other from 1973 to 1990. Both segments of history were times of changing opportunities for oil companies

Several factors distinguish the post-World War II period from the time prior to the war. First, increased demand for oil outside of the United States created a large international oil market and gave an incentive for U.S. oil companies to use their expertise in new geographic markets. Second, the costly experience that companies encountered with long-term oil supply contracts encouraged firms to vertically integrate. Third, the discovery of low-cost oil deposits in the Middle East induced American firms to enter international oil production markets. Fourth, technological changes in the petrochemical industry enabled oil to supplant coal in the production of organic chemicals and provided incentives for oil companies to dominate enter and eventually dominate the petrochemical market. Fifth, the regulatory environment altered incentives by restricting both imports of foreign oil and production of domestic oil. Sixth, the formation of OPEC increased the bargaining power of the oil producing countries relative to the oil producing firms, shifting the economic rents associated with low-cost sources of oil to the producing countries and away from the firms.

Events and technological changes also provided an incentive for firms to enter new lines of business. Many were encouraged to enter the chemical business, particularly fertilizers and petrochemicals. Some firms also perceived advantages in diversifying into uranium and coal production. Other companies took advantage of both depletion allowances, which were tax deductible from domestic profits, and foreign royalty payments in excess of the domestic tax rate, which could be deducted from earnings abroad, to accrue tax-loss carry-forwards that could be applied to the profits of any new acquisition.

The rise and fall of OPEC as a world oil price regulating body dominated events during 1973-90. The direct impact of actions by OPEC leaders on oil companies was the loss of control and ownership of international oil operations, a drop in the number of investment opportunities, and an increase in profits stemming from the price increase and the constant cost of production of domestically produced oil. Managers, federal regulators, and investors reacted to the new circumstances in several ways. First, many oil companies placed a new emphasis on finding domestic sources of oil. Second, seeking to balance international earnings and recognizing that most new sources of domestic oil were already discovered, oil companies diversified into new businesses. Third, before 1980, the government maintained price ceilings on domestic oil coming from wells discovered prior to 1973, mandated crude-sharing relations between producers and refiners, and strictly enforced anti-trust regulations. After 1980, the government lifted controls and relaxed anti-trust enforcement. Fourth, recognizing that excess capacity existed in refining, investors in the highest tax bracket paid 85 percent of cash flow devoted to dividends in taxes, and managers were undertaking unprofitable diversification strategies. Boone Pickens developed royalty trusts as a vehicle to transfer cash flow directly to shareholders at a 70 percent tax rate.

Aspects of the domestic oil industry, the incentives for vertical integration, the regulatory environment, international events, tax policy, and the growth of the petrochemical business are all discussed in more detail in the subsequent appendices.

A.1.2 *Domestic Oil Industry.* Rising U.S. demand for gasoline, heating oil, railroad fuel, and airline fuel led to an increase in U.S. oil consumption from 2.375 billion barrels in 1945 to 4.788 billion barrels in 1968. Concurrent with the increase in consumption was the decline of the United States as the world's major source of oil. U.S. imports rose from 114 million barrels in 1945 to 1.038 billion barrels in 1968 (table A.1.1), or about 20 percent of total U.S. demand. Imports continued to rise throughout the 1970's and reached 2.41 billion barrels of oil, or 40 percent of the domestic market by 1987.[1]

Refiners in the post-World War II period shifted to larger refining plants. Average refinery size rose from 32,000 barrels per day in 1951 to 113,000 barrels per day in 1968 and the number of firms processing at least 50,000 barrels of oil per day rose from 20 in 1955 to 39 in 1975 (Markun, 1976). As refinery size grew, however, the number of independent refiners dropped from 218 in 1950 to 139 in 1968 (Shaffer, 1968).

---

[1]See Table 1, Section IX , *Basic Petroleum Data Book*, 1988, Volume VIII, Number 3.

A greater change in the industry came from OPEC's influence over oil prices. Producers of domestic oil realized a gain in profitability as domestic prices rose in response to world prices. Other companies, however, were less fortunate. Refining companies, for example, encountered risky supply conditions from overseas producers, while nationalizations abroad resulted in a loss of producing properties at other companies. Moreover, for all companies with domestic refining capacity, the increase in oil prices resulted in reduced demand as total U.S. consumption dropped from a high of 9.89 billion barrels in 1978 to 7.56 billion barrels in 1983 and 8.40 billion barrels in 1987 (Shaffer, 1968).

High oil prices and excess capacity in refining led investors to undervalue company assets in the stock market. Chairman Boone Pickens of Mesa Petroleum, recognizing lagging stock prices and wishing to enhance firm value, extended the concept of production payments through his establishment of Royalty Trusts at Mesa in 1979.[2] Royalty Trusts are similar to production payments in that shareholders become direct partners in the ownership of a production field and revenues after expenses go directly to them. By bypassing management completely, Royalty Trusts pay no corporate tax and thus allow shareholders to double their potential dividends. Moreover, it eliminated management discretion over the use of cash flow.

To make Royalty Trusts functional, Pickens spun off a new entity that owned a share of Mesa's oil and gas reserves while maintaining the remainder of the organization as a production company. Investors reacted favorably to the innovation by bidding up stock prices of the new entity to a value worth 90 percent of assets while at the same time valuing major integrated oil companies at about 40 percent of assets (Businessweek, 1983a).

Based on his success with Royalty Trusts at Mesa, Pickens embarked on a program of buying shares of stock in larger companies and subsequently forcing either mergers or restructurings of the companies. Firms that he targeted included General American Oil, Supron Energy, Cities Service, Superior Oil, Gulf, UNOCAL, and Phillips; all of which, except the last two, were taken over by other companies. The industry reacted to the new source of pressure through stock repurchases and dividend payments (table 8.3).

A.1.3 *Vertical Integration.* Teece (1976) describes an integrated oil firm as an oil-refiner in a high-volume business in which high capacity utilization is necessary

---

[2]Production payments are agreements between a production company and private investors in which investors borrow money from a bank and use the proceeds to acquire the rights to a share of the revenue from the production field. Because the production field has proven reserves, the bank uses the contractural agreement between the company and the investor as a form of low-risk collateral and thus extends the loan. Managers at several oil companies, including Mobil, Texaco, and ARCO, used these agreements during the 1970's as a way to help finance their oil development programs.

for competitive survival. He asserts that both reliable sources of crude oil and immediate markets for output are essential for achieving this end.

Low-cost refinery operation requires a processing plant in which trade-offs are made between refinery flexibility and construction costs. Anderson (1984) indicates that since sulphur is a corrosive agent and crude oil sulphur content varies significantly among oil sources, sulphur content must be compatible with the processing equipment. Further, he suggests that construction costs rise as processing flexibility rises.[3] Hence refineries built for low-sulfur content oil are the least costly to build but since they are dependent on a narrower range of crude oil feedstocks, they are more likely to invite opportunistic behavior by a crude oil supplier.

Not only must refineries be compatible with the crude oil input, but their rated capacities must be designed to minimize both transportation and processing costs. In the absence of transportation costs, a large refinery is a low cost producer, making a single refinery optimal.[4] Alternatively, if no production costs do not exist, there would be one refinery per customer. But since both transportation and production costs exist, firms must consider both when designing a refinery. Anderson (1984) reports that considerations of transportation costs, production costs, and refinery flexibility resulted in refinery sizes varying from 2,000 to 500,000 barrels in capacity.

Several economists (Teece, 1976; Mitchell, 1976; Mancke, 1976) assert that low-cost refinery operation may require vertical integration. They claim that when investments are highly specific under conditions of "small numbers bargaining", market solutions may fail. Consequently, since refiners make large investments in a highly specific asset, the refinery, that requires continuous throughput for full capacity utilization, they may be subject to either an opportunistic supplier or an opportunistic marketer. Under such conditions, a supplier, for example, can demand a price equal to the price of refined oil minus the average variable cost of refinery operation. As a result, the refiner cannot recover his sunk costs.

Backward integration into production offers two advantages. First, it eliminates the possibility of opportunism by a supplier. Second, it allows a refiner to build a low-cost refinery that is designed with a limited amount of flexibility. Complete backward integration, however, may not be necessary if a refinery has access to multiple sources of a single-grade oil. As indicated in table A.1.2, most oil firms were not completely backward integrated.

---

[3]Refineries able to process high-sulphur crude can also produce low-sulphur crude, but not vice-versa.

[4]Prindle (1981) reports that there are increasing returns to scale in refinery size.

Teece (1976) points out that pipelines can have competition from barges and other pipelines and also require a large fixed investment, making them susceptible to an opportunistic refiner. Alternatively, there may be one pipeline and many refineries, making a refinery subject to an opportunistic pipeline owner. As a result, there is an incentive in some cases for either backward integration into supply pipelines or forward integration into discharge pipelines. Teece (1976) reports that most oil firms partially integrated into pipelines.

A secure supply and transportation network does not guarantee that a marketer exists to sell the final product. Under conditions when there are multiple refiners but only one distributor, any single refiner may be subject to an opportunistic distributor. Alternatively, if there is one supplier and many distributors, any one distributor can be subject to opportunistic behavior. Accordingly, firms in some markets will find it advantageous to forward integrate into distribution.

Johnson (1976) suggests that Standard Oil of New Jersey, Marathon, Mobil, Standard of Ohio, and ARCO all integrated their operations after they were divested from the Standard Oil Trust in 1911. He adds that those that did not integrate soon disappeared. Moreover, as indicated in table A.1.3, most oil companies had at least partially integrated by 1975. Mitchell (1976) reports that integrated operations allowed those firms to operate refineries at 82 percent capacity, while nonintegrated firms operated their refineries at 55 percent.

Teece (1976) argues that vertical integration may not be necessary under three conditions. First, if multiple sources of oil (marketers of products) exist, then competitive conditions are present and a refiner may not be subject to opportunistic behavior. Second, a refiner may have a locked-in source of supply or market for his output. For example, locally produced crude may be supplied to a refinery adjacent to a utility, enabling a nonintegrated refiner to purchase crude at a price of the spot price minus the shipment cost of the producer to an alternative market. Third, if both the supplier (distributor) and refiner make highly specific investments in a "small numbers" environment, then neither can act opportunistically and a competitive solution obtains.

Anecdotal evidence from the oil industry during the 1970's appears to support Teece (1976). Gulf Oil, for example, lost many of its overseas oil properties after several OPEC countries undertook oil property nationalization programs in the 1970's. Severed from its traditional sources of oil, Gulf responded to its new condition by consolidating its refining and marketing operations (Minard, 1977). Similarly, Texaco had invested in a network of small volume, inflexible refineries that were optimum for conditions of ample supplies of high quality crude and high transport costs. After 1970, however, Texaco faced an environment in which its holdings of high quality crude oil were rapidly being depleted and transportation costs were dropping. As a result, Texaco had to pay a premium on the market for sufficient quality crude oil to supply its refineries while its transportation cost advantage over its competitors dissipated (Cook, 1978).

A.1.4 *Regulatory and Tax Environment.* Regulation of the oil industry during 1930-73 was designed to protect the oil industry from changes in supply conditions brought about by the discovery of low-cost sources of crude. Prorationing laws established production quotas for producers in their states both as a way to more efficiently remove oil from a common resource pool and to protect existing producers from lower cost domestic sources. Oil Import Quotas were imposed by President Eisenhower for the purpose of reducing imports of low-cost foreign crude oil.

During 1973-90, federally mandated legislation directly influenced oil company profitability and impacted the industry. The buy/sell program instituted by the federal government compelled crude surplus firms to share oil with crude-deficient firms, while price controls acted to limit the profitability of domestic produced oil. More-over, anti-trust enforcement and the threat of divestiture limited horizontal investment alternatives.

Tax laws may have also affected the locus of firm investment activity. United States tax laws require that foreign source income be taxed at a rate no less than the United States tax rate. In a 1950 ruling, however, tax payments made in excess of the United States tax rate in one country could be used to offset taxes from other countries that are less than the United States rate. Therefore, incentives existed for firms to enter low-tax rate foreign markets in order to use all available tax credits.

A.1.4.1 Prorationing Laws. The discovery of low-cost oil in eastern Texas in the late 1920's flooded the domestic market with oil and reduced high grade oil from $1.10 per barrel to $0.25 per barrel. The major oil companies tried to set a price at $0.67 per barrel but independent refiners ignored the floor, buying oil for as low as $0.10 per barrel. Having failed to establish a collusive agreement on prices, the oil compa-nies lobbied the state legislature for production regulation.[5]

Controlled production could not only prevent the bankruptcy of higher cost producers but it could also allow a more efficient exploitation of oil reserves. Under conditions of common resource pooling and a "law of capture" legal framework, competitive behavior can lead to an output greater than that compatible with the natural propulsion mechanism existing in the geological formation. By reducing the dis-charge rate, however, inefficient exploitation can be prevented.

Discharging oil from a subterranean pool requires that oil in porous rock migrate to the pump. Much of the oil, however, is trapped in small pockets that have very narrow channels leading to the main avenue of discharge. If another substance, water

---

[5]See Prindle (1981) for a detailed discussion on early Texas prorationing laws.

for example, seeps into the smaller channels or floods the oil avenue before all of the oil has been discharged, then the oil remains locked in the smaller pockets.

One way to limit the losses is to pump at a lower rate. Accordingly, a profit maximizing owner of an entire field will set a rate compatible with the most efficient rate. However, when firms own only a fraction of a field and have leaseholds adjacent to others, they can maximize output by sinking wells near the boundaries of their leaseholds and inducing oil under the adjacent tract to migrate to their own pumps. As a result, if all leaseholders act identically, an excessive number of wells are sunk and over-pumping occurs.

Under certain conditions, voluntary prorationing can be used in order to limit efficiency losses. In the West Texas fields, for example, there were a few large producers who reduced output through voluntary agreement to the maximum efficiency rate. But there were 12,000 producers of various sizes in the East Texas fields, making voluntary compliance unlikely and government intervention necessary.

After lobbying by integrated oil firms and large independent producers, the Texas Railroad Commission was given the responsibility of regulating production. However, enforcement was not effective because small independent producers circumvented state laws by shipping oil across state lines. After the large oil firms lobbied Congress, however, the "Connally Hot Oil Act" was enacted, making the transportation of illegally produced oil across state lines a federal offense.

The economic interests of large and small producers conflict over spacing regulations. Large producers want wide spacing and early unitization of the field as a way to maximize total field output.[6] Small producers, on the other hand, desire closer spacing of wells and delayed unitization as a way to maximize their own production.

The regulators favored small producers in two ways. First, each producer could pump at least ten barrels of oil per day. Second, each owner was allowed at least one well and was allowed an advantageous spacing for others. Moreover, owners of tracts of land that abutted others were permitted to locate one pump equidistant from the pump of his competitor as a way to limit oil migration.

Because the Texas Railroad Commission regulated 50 percent of domestic oil production, it was able to affect the price of crude oil. The commission set prices by first calculating the amount of oil supplied by other areas of the country and then deducting that from total demand to establish an output rate in Texas. Each producer was then limited to producing only on a certain number of days.

---

[6]Unitization refers to single management of all oil field properties rather than independent management by each producer.

Regulation both limited output (table A.2.4) and raised prices. Not once in the period between 1945 and 1972 did the production ceiling reach 100 percent of theoretical maximum. Burrows and Domencich (1970) contend that this restraint added about $1.00 per barrel to the price of oil. As a result, production constraints guaranteed survival to marginal producers while providing an economic rent to low-cost producers.

Regulation of the oil fields resulted in higher crude oil prices and encouraged refiners to find low-cost sources of oil. In the 1950's, low-cost oil from the Middle East began to arrive on the U.S. market. The imported oil and competition from other states, such as Louisiana, forced the Texas Railroad Commission to further reduce output. As a result, oil output in Texas dropped to 28 percent of capacity in 1964 as oil imports rose from 310 million barrels in 1950 to 777 million in 1963 and 1.038 billion in 1968. During that period, the market power of the Texas Railroad Commission diminished as Texas shrank from having the largest known oil field in the world in 1950 to the 24th largest in 1969, about one-tenth the size of the largest field in Kuwait.

A.1.4.2 Oil Import Controls. As low-cost Middle Eastern oil made its way onto the American market in the mid-1950's, it threatened the existence of thousands of marginal producers. In response, President Eisenhower introduced a voluntary compliance program to restrict imports. The only firms, however, that adhered to the quotas were the large international oil firms. As a result, the market share of imports held by small producers rose from 33 percent in 1954 to 55 percent in 1958 and the number of American international producers rose from 11 to 55.

Having failed to stem oil imports with voluntary controls, President Eisenhower instituted a mandatory oil import program in 1959, the Oil Import Quota. The main thrust of the policy was to limit imports to about nine percent of total demand in the area east of the Rockies and restrict quantity to a level sufficient to balance demand in the West. As a result, crude oil brought a price of $3.00 in the U.S., about $1.25 greater than its production and transportation cost (Burrows and Domencich; 1970). There were five main provisions of the Oil Import Quota. First, crude oil could be imported by any refinery at a rate which granted smaller refiners proportionately more imported oil per unit of rated capacity. Second, only firms that imported finished products in 1958 were permitted to import them during the 1960's. Third, residual oil was permitted to enter with no restrictions. Fourth, Canada was permitted unlimited exports to the U.S. market. Fifth, firms that produced petrochemicals were allowed to import a sufficient quantity of oil to meet processing needs.

Relative to the voluntary program, the Oil Import Quota benefited large firms in two ways. First, it guaranteed them a fixed share of the crude import market,

---

[9]The cost varied from $40,000 to $2 million in Romano's survey.

preventing their market share from eroding. Second, it froze their historically high share of the imported final product market at 1958 rates.[7]

To reduce oil costs, coastal and interior refiners could negotiate a mutually beneficial trade of oil import licenses. Coastal refiners could import crude at cost plus a tanker fee, while interior refiners had to pay an additional transportation fee from a coastal port to their refinery. Alternatively, the coastal firm had to use domestic oil that included a transportation cost from a more distant oil field. As a result, imported oil relative to domestic oil was more valuable to a coastal refiner, lending itself to a market in import quotas. In a perfectly competitive environment, refiners would agree on a price of oil that was mutually beneficial. Burrows and Domencich (1970) and Anderson (1984), however, assert that perfect competition did not exist and that coastal refineries benefited proportionately more than interior refiners.

A.1.4.3 Federal Crude-Sharing and Price Control Regulations. After OPEC acted to threaten an embargo of oil exported to the United States in 1973, Congress enacted the Emergency Petroleum Allocation Act of 1973 (EPAA) as a way to limit the oil supply risk faced by refiners of the high-cost imported oil. Under this program, companies with a crude surplus were forced to redistribute oil to companies with a crude deficit at a price set at the average acquisition price of the 15 largest companies in the industry. After enactment of the legislation, however, the price of crude oil leveled off until late 1978, as the marginal cost of oil approached average cost.[8] Hence, little use of the program was necessary until the second price increase in 1979.

The most significant impact of the program came from late 1978 to early 1981 when 215 million barrels of oil were redistributed and the price difference between Saudi Arabian and Iranian oil increased from $0.50 per barrel in January 1979 to $13.65 in December of 1979 (Krapels and others, 1984). In the absence of regulation, companies with a Saudi Arabian supply source of oil could have earned far greater profits while companies with an Iranian supply source may have been forced out of existence. Krapels and others (1984) reported that the net subsidy paid to refiners and consumers by crude-rich companies amounted to approximately $1.5 billion. A change in federal administration in 1981, however, resulted in a discontinuation of the program.

Aside from regulating oil supplies, Congress also initiated a multi-tiered price control program over domestically produced oil from 1974 to 1981. Under this legislation, oil produced from sources discovered before 1973 were priced at a lower cost than newly discovered oil. Krapels and others (1984) note that in January of 1979,

---

[7]See Shaffer (1971), pages 179 and 201 for supporting data.

[8]For a broader discussion of the effects of this legislation, see pages 58-59 of "The Domestic Refining Industry: Economics and Regulation" in *Oil Shock* by Edward N. Krapels, William Colglazier, Barbara Kates-Garnick, and Robert J. Weiner.

the difference between domestically produced oil and imported oil amounted to $4.48 per barrel and rose to $10.07 per barrel by March of 1980. As it did with the buy/sell program, the Reagan administration allowed price controls to expire in 1981.

A.1.4.4 Anti-Trust Policy. Consolidation in the oil industry has been a trend since its early history. Prior to the 1980's, however, merger agreements were generally regional companies operating in different markets. For example, Atlantic Refining, which operated on the East Coast, merged with Richfield, a West Coast company, to form Atlantic-Richfield (ARCO) and Union Oil, a West Coast company, acquired Pure Oil a midwestern firm.

Anti-trust enforcement effort first increased and then decreased after 1970. Immediately following the price increases of the early 1970's, Congress initiated legislation to force divestiture of the large integrated oil companies. Although legislation failed to make it out of Congress, a vote on the issue did occur in 1975. After failure of this initiative, however, anti-trust enforcement and divestiture appeared to weaken. By 1981, anti-trust sentiment had relaxed to such an extent as to allow a merger between DuPont and CONOCO and later mergers between Gulf and Chevron and between other companies. Speaking in an August 1981 interview, Justice Department Anti-Trust Chief William Baxter indicated that by August 10, 1981 one-sixth as many anti-merger suits were filed compared to a similar period during the previous administration (*Fortune,* 1981a). Moreover, he suggested that vertical integrations, such as that between CONOCO and DuPont, would not be discouraged, while mergers between Mobil and CONOCO may be allowed if there was also a divestiture program.

A.1.4.5 Federal Tax Laws. Wolfson and Sholes (1989) and Wolfson (1985) show that tax incentives affect institutional choices. For example, Wolfson (1985) found evidence that both the sharing rules in limited partnership oil and gas programs and the choice of the partnership form over the corporate form were affected by tax considerations. Wolfson and Sholes (1989), on the other hand, linked the increase in merger activity in the 1980's to changes in the tax code.

During the 1950-70 period, assorted tax breaks for depletion and investments were granted. They did not, however, favor the multidivisional form over the functional form. Rather, they encouraged a higher rate of investment and oil exploration.

One tax break granted by the Treasury Department to international firms may have affected the locus of investment activity. In the late 1940's, first Venezuela and then the major Middle Eastern countries imposed a 50 percent royalty on the gross profits of each barrel of crude oil produced in their countries (U.S. House of Representatives, 1974). The Treasury Department, in a 1950 ruling, allowed these payments to be considered as taxes and averaged with taxes in other countries. Since

U.S. taxes must be paid on foreign income that is taxed at a lower rate than the U.S. tax rate, tax averaging provided firms with existing tax credits from high-tax countries an incentive to invest in countries with low tax rates as a way to use up surplus credits. Moreover, the Treasury Department allowed tax credits not used in one year to be carried forward for up to five years. The *Congressional Quarterly* reported that in 1974 these tax credit carry-forwards amounted to about $2 billion.

A.1.4.6 State Tax Laws. State incorporation laws can affect the choice of corporate organizational arrangements but can do so only if the cost of changing the state of incorporation is greater than the benefits. According to Romano (1985), competition among states for the revenues obtained from granting state charters and the minimal costs of changing charters did produce incorporation laws that were favorable to the firm.[9] Delaware, for example, does not differentiate between alternative corporate forms. Moreover, even in states that did draw distinctions, such as New York, incorporation laws may not have affected organizational form. Chandler (1966), for example, indicates that the New York tax code did induce Exxon to adopt a holding company form in 1926 but the company retained centralized control over subsidiaries that were structured as divisions. Therefore, since only Murphy Oil changed states of incorporation between 1950 and 1975, state incorporation laws do not appear to be significant (*Moody's Industrial Manual*, various issues).

A.1.5 *International Oil.* The post-World War II period saw the rest of the world eclipse the U.S. as both the major source of crude oil its largest consumer (table A.1.5). Between 1950 and 1968, for example, world output of crude oil rose from 3.803 to 14.088 billion barrels per year while the U.S. share of output dropped from 51.7 percent to 23.7 percent (Adelman, 1972). Moreover, U.S. oil reserves stagnated at 31 billion barrels after 1954, while reserves rose from 73.6 billion in 1949 to 382 billion in 1967 (Burrows and Domencich, 1970).

Most of the added production did not go to the U.S. In 1957, the U.S. consumed about 3.21 billion barrels of oil per year, while the rest of the world used about 2.45 billion barrels per year. By 1967, however, U.S. consumption had increased to 5.58 billion barrels per year, while that of the rest of the world had risen to 6.32 billion per year. The main source of new demand came from Europe, where oil consumption rose from 0.62 billion barrels per year in 1953 to 2.95 billion barrels per year in 1966 (table A.1.6).

The large integrated international firms were the firms best positioned to supply crude to the European market. By 1954, each of them was participating in at least one production agreement with a major Middle Eastern producing country on oil tracts that contained 90 percent of the oil reserves in the region. The Iranian Consortium, the last one formed, took its final shape in 1953 after the Iranian government dissolved

---

[9]The cost varied from $40,000 to $2 million in Romano's survey.

its 100 percent agreement with British Petroleum. Other agreements included Exxon, Texaco, Chevron, and Mobil with Saudi Arabia in 1948 and Gulf and British Petroleum in 1934, each with a 50-percent stake in Kuwait (table A.1.7). In spite of the dominance of the major international firms on the most productive tracts, profits of up to $1.18 per barrel by 1960 gave sufficient incentive to many independent producers to begin operations in the Middle East, increasing their number from 35 in 1953 to 350 in 1972.[10]

Sampson (1975) argues that actions by oil companies with Middle Eastern oil operations to lower prices and thereby lower royalty payments in 1960 prompted the major oil producing countries to form the Organization of Petroleum Exporting Countries (OPEC). He suggests that OPEC at this time did not act to raise prices but by acting in unison did prevent further reductions. In 1969, however, Colonel Muammar el-Qaddafi of Libya, acting independent of OPEC, compelled Occidental, a company dependent on Libya for most of its oil, to raise its oil prices by 30 percent, pay a $0.30 per barrel tax, and provide a 58 percent royalty increase. Subsequently, other Middle Eastern countries followed. Later, Qaddafi raised prices another $0.91 per barrel.

OPEC more strongly asserted itself in August 1973, when it acted to raise oil prices above the level that existed on the spot market. Prices rose even further in October when an oil embargo placed on the United States because of its support of Israel caused the price of Nigerian crude to rise to $16 per barrel (Anderson and Boyd, 1983). By 1974, however, the price of crude stabilized at approximately $11 per barrel for Arabian light and only rose to $13 per barrel by late 1978.[11]

The 1979 overthrow of the Shah of Iran, a stable ally of the United States, and the outbreak of the Iran-Iraq war in 1980 caused Arab light oil to double in price to a high of $34 per barrel by November 1, 1981. In response to the price increase, oil consuming countries undertook conservation efforts which, combined with a recession, led to a 26-percent decline in U.S. oil consumption by 1983. Moreover, the high price of oil encouraged exploration and expanded production in non-OPEC countries. As a consequence, the price of oil dropped dramatically, until it reached $17 per barrel in 1987.

A.1.6 *Petrochemicals.* Beginning in the mid-1800's, synthetic organic chemicals were made from coal-tar, but petroleum displaced coal in many markets by 1920, resulting in about 21 million pounds of non-coal-tar synthetic organic chemical

---

[10]See *Vertical Integration in the Oil Industry* by Richard Mancke for information on the profitability of Middle Eastern oil and Anderson (1984) about entry into the Middle Eastern market.

[11]Oil prices for this chapter can be found in *Basic Petroleum Data Book: Petroleum Industry Standards*, Volume VIII, Number 3, Section IV (Table 11), Section VI (Table 11), and Section IX (Table 1).

production. Over the next two decades, technological changes in both petrochemicals and organic-chemical-based synthetics led to an increase in demand to over 3 billion pounds by 1939, almost ten times greater than that of coal-tar synthetics (Haynes, 1954). Further developments of synthetics during World War II led to technological advances in the development of petroleum-based substitutes used in the production of rubber, explosives, textile, and drug products. After World War II, demand for petrochemical-based products continued to rise.

Anderson (1984) asserts that oil companies are the natural producers of petrochemical feedstocks and intermediates. He indicates that naphtha, natural gas, and the refinery gases of ethane, propane, and butane are the basic building blocks of petrochemicals, while the primary intermediates include ethylene, propylene, butadiene, benzene, and paraxylene. Moreover, Williams Haynes (1954) reported that since refinery discharge gases are use in the production of petrochemicals, Carbide and Carbon Company in 1935 decided to locate a petrochemical complex processing plant adjacent to an Amoco refinery in order to use the discharge gases as inputs to its petrochemical process. Later, Carbide and Carbon struck a similar deal with Pan American on the Gulf Coast.

Not only were refinery by-products useful for petrochemical production, but strong technological similarities exist between petrochemical production and oil refining. For example, two chemical processes are necessary for the production of refined petroleum products. First, superfractionating processes decompose natural hydrocarbons into their pure components and are necessary for producing gasoline additives. Second, polymerization, a way of constructing more complex hydrocarbons, is used in the production of lubricating oil.

The technological similarities between petroleum products and organic chemicals resulted in several important chemical discoveries by oil firms. Standard Oil of New Jersey, for example, developed butadiene, while experimenting for additives to lubricating oil. Later, after World War II, Exxon extended its research efforts into agricultural chemicals, plastics, detergents, and other materials. Shell, Chevron, and Texaco were likewise early developers of petrochemicals. Moreover, Chandler (1966) indicates that by the mid-1950's Amoco, Gulf, Sinclair, Mobil, Phillips, CONOCO, and Standard of Ohio had also entered the petrochemical business.

A.1.7 *Coal.* Coal and oil production have technological similarities, making the coal industry a possible related investment opportunity for oil companies. Speaking before the Senate Interior Special Sub-committee on Integrated Oil Operations on December 6, 1973, Mr. Randall Meyer, president of Exxon, suggested that coal, like petroleum, is capital intensive, serves a similar market, and benefits from the hydrocarbon processing and pipeline expertise of oil companies. Since physical operations are not redundant, however, few cost savings can be obtained through a consolidation of coal and petroleum operations.

Neither rapidly rising prices nor rising demand appear to offer a clear incentive to enter the coal market. United States energy consumption shifted away from coal after World War II as natural gas and oil replaced coal for powering railroads and heating buildings. As a result, demand for coal dropped from 545,000 tons in 1947 to 366,000 tons in 1958. Afterward, electricity demand rose and coal demand increased to 550,000 tons by 1974 (Zimmerman, 1981). During that time, prices in 1967 dollars rose from $0.749 per million BTU in 1947 to $1.046 in 1958 and $3.332 in 1975.

A.1.8 *Uranium.* After World War II, uranium demand by the United States Atomic Energy Commission (AEC) created a market in a new energy commodity. To encourage production, the AEC followed a cost plus mark-up pricing strategy. Producers responded with an increase in output from almost zero in 1950 to 17,640 tons in 1960 (Owen, 1985). After filling stockpiles, however, AEC demand dropped, resulting in a decline in production to a level of about 10,000 tons annually from 1964 to 1975.

Uranium profits appeared to be quite high initially. The AEC paid prices ranging from $8.53 to $12.51 per pound between 1950 and 1962, but then set a flat rate at $8.00 between 1962 and 1967. With average costs of only about $5.00 per pound, profits in excess of $3.00 per pound were possible (Owen, 1985).

Oil firms that entered the uranium exploration market may have been attracted by the high prices offered by the government and the similarities that uranium exploration had with oil exploration. Ezell (1979) reported that Kerr-McGee entered the uranium prospecting business because it was similar to oil exploration in that both exist in sedimentary rock in remote areas. Further, Randall Meyer of Exxon, speaking before the Senate Interior Special Sub-Committee on Integrated Oil Operations in 1973, asserted that similarities included: capital intensity, remoteness of operations, long lead times from discovery to final product, and the rock structures in which uranium exists.

A.2 Summary and Discussion
Developments in the domestic oil market, technological change, and domestic resource depletion encouraged firms to adapt their domestic operations in two ways. First, improvements in refinery technology and a growing homogenization of the domestic market made large refineries a low-cost alternative to smaller ones.[12] Second, larger refineries and the need to run them at near full capacity made a guaranteed source of supply important and motivated firms to vertically integrate.

---

[12]Chandler (1966) reports that the oil market was becoming more homogenous in the period following World War II.

The Oil Import Quota also offered several incentives to firms. First, it encouraged mergers between coastal refineries and interior refineries because, as a single firm, the coastal refinery could use all import allowances and the interior refinery could use only domestic oil. Second, it provided an incentive for an East Coast refiner to purchase a West Coast company in order to import and process crude on the West Coast and ship refined products to the East Coast. Third, it discouraged American producers with no market outlets overseas from international exploration, while it encouraged those with existing foreign production to either build foreign marketing outlets or to sell their foreign interests. Shaffer (1968) argues that many firms increased their foreign marketing activities.[13] Fourth, since residual oil imports were not controlled, refiners converted their product mix to an output with proportionately more gasoline and imported residual. Fifth, firms that had foreign residual production but no import licenses were encouraged to merge with residual oil distributors. Shaffer (1968) indicates that Texaco, Chevron, and Shell all increased residual oil allocations in this way. Sixth, it encouraged exploration in the United States and Canada at the expense of additional imports.

Prorationing laws also affected incentives. These laws encouraged domestic exploration in states not imposing production restrictions in order to realize high domestic oil prices while producing at oil-well capacity.

Four factors provided oil firms with an incentive to search for Middle Eastern oil. First, Burrows and Domencich (1970) estimate that the long-run cost of extraction, exploration, and production of oil in the Middle East was about $0.15 to $0.30 per barrel, while costs in the U.S. ranged up to $3.00 per barrel. Second, because of prorationing laws and import restrictions, the American cost of crude was approximately $3.00 per barrel, while the cost of imported oil, including transportation and royalty costs, was approximately $1.75 per barrel. Third, table A.1.6 indicates a large increase in demand outside the U.S., as both Europe and the rest of the world began to recover from World War II. Fourth, American tax policy allowed royalty payments made by a producing firm in excess of the domestic tax rate to be deducted from taxes owed from countries with lower tax rates.

Many incentives existed for diversification into both related and nonrelated businesses. Foreign tax credits, for example, gave firms large tax-loss carry-forwards that could be applied to foreign earnings from oil or non-oil businesses. Further, the formation of OPEC made Middle Eastern oil a more risky business, giving firms an incentive to both diversify production properties geographically and diversify into other businesses as a hedge against the long-term viability of the oil business (Anderson, 1984). Moreover, oil firms generated large cash flows from oil sales. This income could be used for either investments into other businesses or as dividends to

---

[13]Shaffer points out that total investment, largely for marketing development, rose from 23 to 30 percent of total investment between 1959 and 1968.

stockholders. Finally, federal anti-trust regulations acted to discourage horizontal mergers, while EEPA encouraged excess refining capacity by forcing crude-rich companies to share oil with crude-poor firms.

Some new product markets appeared to offer profitable incentives to the oil industry. Petrochemical production, for example, draws refinery gases as feedstocks and is technologically similar to oil refining, making it an attractive market to oil refiners. Another market, uranium, was an infant industry in 1950 and had several similarities to oil exploration. Moreover, producers were initially guaranteed costs plus a mark-up and it may have been a highly profitable opportunity. Other product markets, such as coal, also have technological similarities to oil, but real cost savings by combination were not apparent. A complete list of markets in which oil firms had entered by 1975 is shown in table A.1.8.

TABLE A.1.1–U.S. PRODUCTION, CONSUMPTION, AND IMPORTS
OF LIQUID PETROLEUM, 1945-68

Year	Domestic Crude Oil Production (billion barrels)	Crude Imports (billion barrels)	Domestic Consumption (billion barrels)	Crude Exports (billion barrels)
1945	1.826	0.114	1.773	0.182
1946	1.852	0.139	1.793	0.153
1947	1.990	0.159	1.990	0.164
1948	2.167	0.188	2.114	0.134
1949	1.999	0.235	2.118	0.119
1950	2.156	0.310	2.375	0.111
1951	2.453	0.308	2.570	0.154
1952	2.514	0.348	2.664	0.158
1953	2.596	0.377	2.775	0.147
1954	2.568	0.384	2.832	0.130
1955	2.766	0.455	3.088	0.134
1956	2.911	0.525	3.213	0.157
1957	2.912	0.575	3.219	0.207
1958	2.740	0.619	3.309	0.101
1959	2.896	0.620	3.450	0.077
1960	2.915	0.664	3.536	0.074
1961	2.984	0.700	3.579	0.064
1962	3.049	0.760	3.736	0.061
1963	3.152	0.775	3.851	0.076
1964	3.209	0.777	3.959	0.074
1965	3.290	0.911	5.125	0.068
1966	3.508	0.939	4.325	0.071
1967	3.730	0.925	4.584	0.112
1968	3.879	1.038	4.788	0.085

Source: U.S. Bureau of Mines. 1968. "Crude Petroleum, Petroleum Products, and Natural Gas Liquids." *Mineral Industry Survey.*

TABLE A.1.2–DOMESTIC AND WORLD CRUDE OIL SELF-
SUFFICIENCY RATIOS, 1972 AND 1974

Company	Domestic		Worldwide	
	1972	1974	1972	1974
Amerada Hess	0.21	0.19	0.52	0.27
American Petrofina	0.10	0.14	0.14	0.11
Apco Oil	0.62	0.17	0.62	0.74
Ashland Oil	0.06	0.07	0.13	0.17
ARCO	0.58	0.61	0.85	0.87
Cities Service	0.83	0.92	0.84	0.96
Clark Oil	0.03	0.03	0.02	0.02
CONOCO	0.65	0.68	1.15	0.87
Diamond Shamrock	0.39	0.39	0.39	0.39
Exxon	0.94	0.79	0.97	0.70
Getty Oil	1.63	1.40	1.88	1.72
Gulf	0.73	0.59	1.65	1.38
Kerr-McGee	1.04	0.24	1.11	0.29
Marathon	0.79	0.66	1.59	1.18
Mobil Corp.	0.46	0.51	0.860	1.16
Murphy Oil	0.23	0.17	0.40	0.36
Pasco	n.a.	0.40	n.a.	0.40
Phillips	0.49	0.49	0.60	0.62
Shell Oil	0.65	0.58	0.65	0.58
Skelly	1.26	0.98	1.32	1.08
Chevron	0.61	0.47	1.54	2.07
Amoco	0.51	0.57	0.79	0.81
Stand. Oil of Ohio	0.07	0.09	0.13	0.16
Sun Oil Co.	0.58	0.54	0.80	0.66
Tenneco	n.a.	0.84	n.a.	0.93
Texaco	0.91	0.85	1.36	4.31
UNOCAL	0.78	0.61	0.77	0.82

Note: Total domestic or worldwide production as a percent of total domestic or worldwide refinery runs.
n.a. = not available.
Source: *National Petroleum News Fact Book.* 1973 and 1975. New York: National
Petroleum News.

TABLE A.1.3 –VERTICAL INTEGRATION OF SELECTED REFINING COMPANIES[1]

Company	Refining Output (barrels/day)	Produces Crude Oil	Owns Crude Pipeline	Owns Refined Pipeline	Markets Gasoline
American Petrofina	200,000	x	x	x	x
Kerr-McGee Oil	166,000	x	x	x	x
Commonwealth Oil	161,000			x	x
Union Pacific	152,200	x	x	x	x
Murphy Oil	137,500	x	x	x	x
Koch Industries	109,800	x	x	x	x
Clark Oil	108,000	x	x	x	x
Tenneco	103,000	x	x	x	x
Crown Central	100,000	x	x	x	x
Oil Shale Corp	87,000		x	x	x
Charter Company	85,900	x	x	x	x
Agway Inc	74,500	x			x
Farmland Inc.	73,838	x	x		x
Tesoro	64,000	x	x		x
Pennzoil	62,600	x	x	x	x
Husky Oil	59,000	x	x	x	x
Apco Oil	58,670	x	x	x	x
United Refining	58,000	x	x	x	x
National Co-op Rfg.	54,150	x	x	x	x

[1]"x" indicates that the company has operations in this aspect of oil production.
Note: Excluded from this sample are the largest companies, all of which are integrated to some degree. These companies include: Amerada Hess, Ashland Oil, ARCO, Cities Service, Coastal States Gas, CONOCO, Exxon, Getty Oil, Gulf, Marathon, Mobil Corp., Phillips, Shell Oil, Chevron, AMOCO, Standard of Ohio, Sun Oil Co., Texaco, and UNOCAL.
Source: Pierson, Walter R. 1976. in *Witness for Oil*, edited by Patricia Malony Markum. Washington D.C.: American Petroleum Institute.

TABLE A.1.4–ANNUAL AVERAGE OF MONTHLY MARKET DEMAND FACTORS
FOR TEXAS AND LOUISIANA, 1950-73

Year	Texas	Louisiana
1950	63	n.a.
1951	76	n.a.
1952	71	n.a.
1953	65	90
1954	53	61
1955	53	48
1956	52	42
1957	47	43
1958	33	33
1959	34	34
1960	28	34
1961	28	32
1962	27	33
1963	28	32
1964	28	32
1965	29	33
1966	34	35
1967	41	38
1968	45	42
1969	52	44
1970	72	56
1971	73	73
1972	94	91
1973	100	100

Note: These averages are presented as a percentage of theoretical maximum.
n.a. = not available.

Source: Krueger, Robert. 1978. *The United States and International Oil: A Report for the Federal Energy Administration on U.S. Firms and Government Policy*, Washington D.C.: Government Printing Office.

TABLE A.1.5*–CRUDE OIL PRODUCTION BY COUNTRY, 1950-68

Year	U.S.	Venezuela	Iran	Iraq	Kuwait	Saudi Arabia	USSR	World
1950	1,974	547	242	50	126	200	266	3,803
1951	2,248	622	124	65	205	278	285	4,283
1952	2,290	660	8	141	273	302	341	4,531
1953	2,357	644	9	210	315	308	380	4,798
1954	2,315	692	21	228	353	348	427	5,017
1955	2,484	787	121	251	407	352	510	5,626
1956	2,617	899	197	232	412	361	612	6,124
1957	2,617	1,014	263	163	439	362	718	6,440
1958	2,449	951	301	266	539	370	834	6,608
1959	2,575	1,011	345	311	547	400	946	7,133
1960	2,575	1,042	386	354	644	456	1,079	7,674
1961	2,622	1,066	432	366	655	508	1,212	8,186
1962	2,676	1,168	482	367	758	555	1,360	8,882
1963	2,753	1,186	538	423	920	595	1,540	9,538
1964	2,787	1,242	619	462	906	628	1,643	10,310
1965	2,849	1,268	688	482	924	739	1,786	11,063
1966	3,028	1,230	771	506	983	873	1,922	11,956
1967	3,216	1,290	956	443	987	950	2,103	12,855
1968	3,339	1,311	1,011	549	1,051	1,036	2,251	14,088

Source: Burrows, James and Thomas Domencich, 1970, *An Analysis of the United States Oil Import Quota*, Lexington, Massachusetts: D.C. Heath and Company.

TABLE A.1.6'–POST-WAR CONSUMPTION GROWTH IN THE MAJOR
OIL CONSUMING COUNTRIES, 1950-70

Year	United States	Western Europe	Japan
1950	2.375	n.a.	n.a.
1951	2.570	n.a.	n.a.
1952	2.664	n.a.	n.a.
1953	2.775	0.624	n.a.
1954	2.832	0.724	n.a.
1955	3.088	0.779	n.a.
1956	3.213	0.836	n.a.
1957	3.219	0.852	n.a.
1958	3.309	0.995	n.a.
1959	3.450	1.163	n.a.
1960	3.536	1.355	n.a.
1961	3.579	1.541	0.256
1962	3.736	1.718	0.311
1963	3.851	1.969	0.377
1964	3.959	2.283	0.456
1965	5.125	2.618	0.556
1966	4.325	2.949	0.739
1967	5.585	3.217	0.877
1968	n.a.	3.667	1.036
1969	n.a.	4.078	1.427
1970	n.a.	4.580	1.460

n.a. = not available.

Source: Burrows, James and Thomas Domencich, 1970, *An Analysis of the United States Oil Import Quota.* Lexington, Massachusetts: C.C. Heath and Company.

TABLE A.1.7'–COMPOSITION OF THE KEY OIL CONSORTIUMS OF THE MIDDLE EAST

Firm	Country			
	Iran	Saudi Arabia	Kuwait	Abu Dhabi
Exxon	7.00	11.88	0.00	11.88
Gulf	7.00	0.00	50.00	0.00
Mobil	7.00	11.88	0.00	11.88
Shell	14.00	23.75	0.00	23.75
Chevron	7.00	0.00	0.00	0.00
Texaco	7.00	0.00	0.00	0.00
Other	5.00	0.00	0.00	5.00
Non-U.S.	47.50	47.50	50.00	47.50
Total	100.00	100.00	100.00	100.00

Source: Sampson, Anthony. 1975. *The Seven Sisters: The great oil companies and the world they made.* London: Hodder and Staughten.

TABLE A.1.8[*]–EXTENT OF INDUSTRY DIVERSIFICATION, 1979

Product Market	Multinationals	Other
Petrochemicals	18	12
Plastics	5	2
Asphalts and coatings	5	3
Organic chemicals	10	4
Fertilizer and agricultural chemicals	4	3
Rubber	2	0
Electrical engineering	2	2
Mechanical engineering	1	3
Metal fabrication	0	1
Construction	3	1
Transportation	8	2
Mining	12	7
Real estate and land development	2	8
Agricultural products	1	3
Lumbering	1	1
Technical services	1	0
Office supplies	0	0
Printing and publishing	0	1
Financial services	0	2
Other non-petroleum interests	1	3

[*] Source: Arabinda Ghosh. 1985. *Competition and Diversification in the U.S. Petroleum Industry.* Westport: Quorum Books.

References

Adelman, Morris A. 1972. *The World Petroleum Market*. Baltimore: The Johns Hopkins University Press.

Alchian A., and H. Demsetz. 1972. "Production, information costs, and economic organization." *American Economic Review* 62 (December): 777-95.

American Petroleum Institute. 1988. *Basic Petroleum Data Book: Petroleum Industry Statistics* 8 (3): various tables.

Anderson, Jack and James Boyd. 1983. *Fiasco*. New York: New York Times Book Company.

Anderson, Robert. 1984. *Fundamentals of the Petroleum Industry*. Norman, Oklahoma: University of Oklahoma Press.

Ansoff, I.I. and R. G. Brandenburg. 1971. "A language for organizational design: Part II." *Management Science* 17 (August): 717-731.

Armour H., and D. Teece. 1978. "Organizational structure and economic performance." *BellJournal of Economics* 9: 106-22.

Benston, G.J. 1985. "The validity of profits-structure studies with particular reference to the FTC's line-of-business data." *American Economic Review* 75: 37-67.

Bhargava, Narottam. 1972. "The impact of organizational form on the firm: Experiences of 1920-70." Ph.D. diss., University of Pennsylvania.

Burrows, James and Thomas Domencich. 1970. *An Analysis of the United States Oil Import Quota*. Lexington, Massachusetts: D.C. Heath and Company.

Businessweek. 1965. "Oil is no longer enough." July 3, 48-57.

Businessweek. 1976. "How Sun Company's split personality works." November 8, 72-73.

Businessweek. 1978a. "The New Diversification Oil Game." April 24, 76-78.

Businessweek. 1978b. "Ashland Oil: Getting out of crude as it sheds lackluster operations." September 4, 90-91.

Businessweek. 1980a. "How contrasting strategies made different companies." June 2, 64-74.

Businessweek. 1980b. "Getty: Spending a windfall to diversify far out of oil." June 23, 99-102.

Businessweek. 1980c. "An oil giant's dilemma." August 25, 62-68.

Businessweek. 1981a. "Why mining looks good to big oil." March 30, 46-47.

Businessweek. 1981b. "Energy growth fuels problems for a conglomerate." November 23, 80-89.

Businessweek. 1981c. "Marathon pumps up its takeover defense." November 30, 52-53.

Businessweek. 1982. "Why things aren't going right for Exxon." June 7, 88-90.

Businessweek. 1983a. "Restructuring big oil." November 14, 138-50.

Businessweek. 1983b. "The boardroom drama now playing at Getty Oil." December 12, 29-30.

Businessweek. 1984a. "Why Texaco values Getty at $10 billion." January 23, 34-36.

Businessweek. 1984b. "Why Royal Dutch Shell is betting on the U.S." February 20, 98-100.

Businessweek. 1984c. "Why Gulf lost its fight for its life." March 19, 76-84.

Businessweek. 1985a. "Will Tenneco's Harvester deal turn out to be ' the corporate equivalent of Vietnam?" February 4, 80-81.

Businessweek. 1988. "Ashland just can't seem to leave its checkered past behind." October 31,122-26.

Cable, John and Manfred Dirrenheimer. 1983. "Hierarchies and markets." *International Journal of* Industrial Organization 1: 43-63.

Chandler, A. D. Jr. 1966. *Strategy and Structure.* New York: Doubleday and Company.

Chandler, A. D., Jr. 1977. *The Visible Hand: The Managerial Revolution in American Business.* Cambridge, Massachusetts: Harvard University Press.

Coase, R. H. 1937. "The nature of the firm." *Economica* 4: 386-405.

Cook, James. 1978. "Rebuilding the house that Long built." *Forbes.* April 17, 47-54.

Cook, James. 1985. "Exxon proves that big doesn't mean rigid." *Forbes.* April 29, 66-74.

Dun's. 1970. "Tenneco's careful diversification." December, 27-28.

Ezell, John Samuel. 1979. *Innovation in Energy: The Story of Kerr-McGee.* Norman, Oklahoma: University of Oklahoma Press.

Fisher, F. and J. McGowen. 1983. "On the misuse of accounting rates of return to infer monopoly profits." *American Economic Review.* 73: 82-97.

Flanigan, James. 1979. "Continental Oil: a time for sowing a time for reaping." *Forbes.* March 19, 70-75.

Forbes. 1968a. "Occidental: Lucky like a fox." June 1, 24-32.

Forbes. 1968b. "Back to the well." November 15, 63-65.

Forbes. 1980. "The odds were right." November 24, 134.

Forbes. 1985. "The value players are salivating." January 28, 58-59.

Fortune. 1981a. "Conoco on the block." August 10, 17-18.

Fortune. 1981b. "What's in it for DuPont?" September 7, 64.

Ghosh, Arabinda. 1985. *Competition and Diversification in the United States Petroleum Industry.* Westport, Connecticut: Quorum Books.

Greene, William N. 1985. *Strategies of the Major Oil Companies.* Ann Arbor, Michigan: UMI Research Press.

Harris, B.C. 1983. *Organizations: The Effect on Large Corporations.* Ann Arbor, Michigan: UMI Research Press.

Harris, Marilyn. 1984. "Exxon wants outs of the automated office." *Businessweek.* December 17,39.

Haynes, Williams. 1954. *American Chemical Industry: A History; Volume V: 1930-39.* New York: D. Van Norstrand Company.

Hill, Charles. 1985. "Internal organization and enterprise performance: Some UK evidence." *Managerial and Decision Economics* 6 (4): 10-216.

Jensen, Michael, and W. Meckling. 1976. "Theory of the firm: Managerial behavior, agency costs, and capital structure," *Journal of Financial Economics* 3 (October): 305-60.

Johnson, Arthur M. 1976. "Lessons of the Standard Oil Divestiture." *Vertical Integration in the Oil Industry.* Edward Mitchell (ed). Washington D.C.: American Enterprise Institute.

Keren, M., and D. Levhari. 1979. "The optimum span of control in a pure hierarchy." *Management Science* 25 (November): 1162-72.

Keren, M., and D. Levhari. 1983. "The internal organization of the firm and shape of average costs." *Bell Journal of Economics* 14: 474-86.

Kirkpatrick, David. 1988. "Hammer hits 90! Oxy grows up too." *Fortune.* November 7, 58-64.

Krapels, Edward N., E. William Colglazier, Barbara Kates-Garnick, and Robert J. Weiner. 1984. "The Domestic Refining Industry: Economics and Regulation." *Oil Shock.* Robert Weiner (ed). Cambridge, Massachusetts: Ballinger Publishing.

Krueger, Robert. 1978. *A Report for the Federal Energy Administration on U.S. Firms and Government Policy.* Washington, D.C.: Government Printing Office.

Kumar, M.S. 1985. "Growth, acquisition activity, and firm size: Evidence from the United Kingdom." *Journal of Industrial Economics* 33 (March): 327-38.

Lazear, E. and S. Rosen. 1981. "Rank order tournaments as optimum labor contracts." *Journal of Political Economy* 89: 841-64.

Lindenberg, E.B., and S.A. Ross. 1981. "Tobin's Q Ratio and Industrial Organization." *Journal of Business* 54 (January): 1-32.

Livermore, Shaw. 1935. "The Success of Industrial Mergers." *Quarterly Journal of Economics* 50 (Nov): 94.

Lucas, R. E. 1978. "On the size distribution of firms." *Bell Journal of Economics* 6 (Spring): 250-80.

Lustgarten, Steven and Stavros Thomadakis. 1987. "Mobility barriers and Tobins's Q." *Journal of Business* 60 (4): 519-37.

Mack, Toni. 1988. "How about Texaco?" *Forbes.* August 22, 34.

Maddala, G.S. 1977. *Econometrics.* New York: McGraw-Hill Company.

Mancke, Richard B. 1976. "Competition in the oil industry." *Vertical Integration in the Oil Industry.* Edward Mitchell (ed). Washington D.C.: American Enterprise Institute.

Markun, Patricia Malony (ed). 1976. *Witness for Oil.* Washington D.C.: American Petroleum Institute.

Meyer, Randall. 1973. Testimony before the Senate Interior Special Sub-Committee on Integrated Oil..

Minard, Lawrence. 1977. "Gulf Oil: yesterday's villain." *Forbes.* October 15, 95-100.

Mitchell, Edward, J. 1976. "Capital cost savings of vertical integration." *Vertical Integration in the Oil Industry.* Edward Mitchell (ed). Washington D.C.: American Enterprise Institute.

Montgomery, Cynthia, and Birger Wernerfelt. 1988. "Diversification, Ricardian Rents and Tobin's Q." *Bell Journal of Economics* 19: 623-30.

Moody's Industrial Manual. Various issues, 1930-90. New York: Moody's Investor Service.

National Petroleum News Fact Book. 1973 and 1975. New York: National Petroleum News.

Norman, James, and Barbara Starr. 1985. "The Pickens vice tightens around UNOCAL." *Businessweek.* April 15, 45.

Oi, Walter. 1983. "Heterogeneous firms and the organization of production." Economic Inquiry 21, (April): 147-71.

Ouchi, W. G. 1980. "Markets, bureaucracies, and clans." *Administrative Science Quarterly* 25 (March): 120-42.

Owen, Anthony. 1985. *The Economics of Uranium.* New York: Praeger Publishers.

Prindle, David. 1981. *Petroleum Politics and the Texas Railroad Commission.* Austin: University of Texas Press.

Radner, Roy. 1978. "A behavioral model of cost reduction." *Bell Journal of Economics* 6 (Spring): 196-215.

Romano, Roberta. 1985. "Law as a product: Some pieces of the incorpo-ration policy." *Journal of Law Economics and Organization* 1 (Fall): 225-83.

Rubin, P. 1978. "The theory of the firm and the structure of franchise contract." Journal of Law and Economics 21 (April): 223-33.

Rumelt, Richard P. 1986. *Strategy Structure and Economic Performance.* Boston: Harvard Business School Press.

Salinger, M.A. 1984. "Tobin's Q, unionization and the concentration profit relationship." *Rand Journal of Economics* 15 (Summer): 157-70.

Sampson, Anthony. 1975. *The Seven Sisters: The Great Oil Companies and the World They Made.* London: Hodder and Staughton.

Santry, David, and Timothy Ord. 1981. "What teamed Conoco and Cities Service." *Businessweek.* July 6, 85.

Schipper, K., and R. Thompson. 1983. "Evidence of the capitalized value of merger activity for acquiring firms." *Journal of Financial Economics* 11: 85-119.

Shaffer, Edward H. 1968. *The Oil Import Program of the United States.* New York: Frederick A. Praeger Publishers.

Shavell, S. 1979. "Risk sharing and incentives in the principal and agent relationship." *Bell Journal of Economics* 10 (Spring): 55-73.

Sherman, Stratford. 1989. "Who's in charge at Texaco now?" *Fortune.* January 16, 68-72.

Sherril, Robert. 1983. *The Oil Follies of 1970-1980.* Garden City, New York: Anchor Press/Doubleday.

Siler, Charles. 1987. "Give it to the stockholders." *Businessweek.* September 7, 54-56.

Simon, Ruth. 1985. "Sohio bucks the trend." *Forbes.* June 17, 41-42.

Silver, M., and R.D. Auster. 1969. "Entrepreneurship, profits and limits of firm size." *Journal of Business* 42 (July): 277-81.

Smirlock, T., T. Gilligan, and W. Marshall. 1984. "Tobin's Q and the structure performance relationship." *American Economic Review* 74 (December): 1051-60.

Spence, Hartzell. 1962. *Portrait in Oil.* New York: McGraw-Hill Book Company.

Steer, Peter and John Cable. 1978. "Internal organization and profit: An empirical analysis of large UK companies." *Journal of Industrial Economics* 27: 13-30.

Teece, David J. 1976. "Vertical integration in the U.S. oil industry." *Vertical Integration in the Oil Industry.* Edward Mitchell (ed.). Washington D.C.: American Enterprise Institute.

Teece, David. J. 1981. "Internal organization and economic performance: An empirical analysis of the profitability of principal firms." *Journal of Industrial Economics.* 30: 173-99.

Thompson, R.S. 1981. "Internal organization and profit: A note." *The Journal of Industrial Organization* 30: 201-11.

Ticer, Scott and William Glasgall. 1985. "The Doctor Performing Arco's Radical Surgery." June 3, 64-9.

U.S. Bureau of Mines. 1968. "Crude petroleum, petroleum products, and natural gas liquids." *Mineral Industry Surveys.*

U.S. Department of Energy. 1984. *Performance Profiles of Major Energy Producers.*

U.S. House Report Number 93-1502. 1974. *Energy Tax and Individual Relief Act of 1974.*

Wall Street Journal Index. Various issues, 1958-90. New York: Dow Jones & Company.

Williamson, O.E. 1970. *Corporate Control and Business Behavior; An Inquiry into the Effects of Organizational Form on Enterprise Behavior.* Englewood, New Jersey: Prentice Hall.

Williamson, O. E. 1975. *Market and Hierarchies: Analysis and Anti-*

trust Implications. New York: Free Press.

Williamson, O. E. 1987. *The Economic Institutions of Capitalism.* New York: The Free Press.

Wilson, R. 1978. "Information economies of scale." *Bell Journal of Economics* 6 (Spring) 184-95.

Wolfson, Mark A. 1985. "Empirical evidence of incentive problems and their mitigation in oil and gas tax shelter programs." *Principles and Agents in the Structure of Business.* John Pratt and Richard Zeckhauser (eds.). Boston: Harvard University Press.

Wolfson, Mark A., and Myron Scholes. 1989. "The effects of changes in tax laws on corporate reorganization activity." *Stanford Working Paper.* April: 1-43.

Zimmerman, Marvin, B. 1981. *The U.S. Coal Industry: The Economics of Policy Choices.* Cambridge, Massachusetts: The MIT Press.

Other Readings

Alchian, A. and S. Woodward. 1988. "The firm is dead; long live the firm; A review of Oliver E. Williamson's Economic Institutions of Capitalism." *Journal of Economic Literature* 26 (March): 65-79.

Baker, George B., Michael C. Jensen, and Kevin J. Murphy. 1988. "Compensation and incentives: Practice vs. theory." *Journal of Finance* 43: 593-616.

Barzel, Y. 1982. "Measurement cost and the organization of markets." *Journal of Law and Economics* 25 (April): 27-48.

Baumol, William J. 1967. *Business Behavior Value and Growth.* New York: Harcourt, Brace, & World, Inc.

Beaton, Kendall. 1957. *Enterprise in Oil: A History of Shell in the United States.* New York: Appleton-Century-Crofts, Inc.

Benston, G.J. 1985. "The self-serving management hypothesis: Some evidence." *Journal of Accounting and Economics* 7: 67-84.

Blaubert, Howard. 1975. *The First Hundred Years.* New York: Dell Publishing Company.

Brudney, Victor. 1985. "Corporate governance, agency costs, and the rhetoric of contract." *Columbia Law Review* 85: 1403-44.

Businessweek. 1960a. "Freeing the big boss' hands." July 16, 97-98. *Businessweek.* 1960b. "Standard of Jersey's new plan for Realignment." August 6, 44-54.

Businessweek. 1962. "Country boys step out." May 19, 145-150.

Businessweek. 1966. "The plan that put fire into Sohio's profits." February 12, 106-16.

Businessweek. 1967. "How they won the West and more." January 28, 178-85.

Businessweek. 1971. "Why Indiana Standard grows so well." December 4, 58-59.

Businessweek. 1973. "Phillips gets its growth under control." March 10, 164-66.

Businessweek. 1977a. "What makes Mobil run?" June 13, 80-85.

Businessweek. 1977b. "Gulf Oil: Yesterday's villain." October 15, 95-100.

Businessweek. 1984d. "Tenneco's ships come in." June 18, 72-73.

Buttrick, J. 1952. "The inside contracting system." *Journal of Economic History* 12 (Summer): 205-21.

Byrne, John. 1988. "The Rebel Shaking Up Exxon." *Businessweek.* July 18, 104-11.

Chakravarty, Subrata. 1988. "Fred raises the flag again, but nobody's saluting." *Forbes.* June 13,42-48.

Cook, Don. 1986. "Will Horton have to take a hatchet job to Standard Oil?" *Forbes* May 12, 78-79.

DeAngelo. Harry and Edward M. Rice. 1983. "Antitakeover charter amendments and stockholder wealth." *Journal of Financial Economics* 11: 329-57.

Evans, David S. 1987. "Tests of alternative theories of firm growth."

Journal of Political Economy 95: 657-74

Fama, E. 1980. "Agency problems and the theory of the firm." *Journal of Political Economy* 88 (April): 288-307.

Fama, E., and M. C. Jensen. 1983. "Separation of ownership and control." *Journal of Law .* (June): 301-26.

Forbes. 1975. "Chemical change." November 15, 68.

Giddens, Paul. 1955. *Standard Oil Company (Indiana): Oil Pioneer of the Middle West.* New York: Appleton-Century-Crofts, Inc.

Giebelhaus, August W. 1980. *Business and Government in the Oil Industry: A Case Study of Sun Oil: 1876-1945.* Greenwich, Connecticut: JAI Press Inc.

Forbes. 1979. "Continental Oil: A time for sowing, a time for reaping." March 19, 70-75.

Grossman, Sanford J., and Oliver D. Hart. 1980. "Takeover bids, the free-rider problem, and the theory of the corporation." *Bell Journal of Economics* 11: 42-64.

Harris, Kenneth. 1987. *The Wildcatter: A Portrait of Robert O. Anderson.* New York: Weinfeld & Nicholson.

Haynes, Williams. 1949. *American Chemical Industry: A History; Vol-*

ume V: Company Histories to 1948. New York: D. Van Norstrand Company.

Holmstrom, B. 1979. "Moral hazard and observability." Bell Journal of Economics 10 (Spring): 74-91.

Jensen, Michael. 1986. "Agency costs of free cash flow, corporate finance and takeovers." *American Economic Review* 76 (May): 323-29.

Jensen, Michael, and Jerald Zimmerman. 1985. "Management Compensation and the Managerial Labor Market." *Journal of Accounting and Economics* 7: 3-10. 1948.

Masse, Joseph. 1960. *Blazer of Ashland Oil.* Lexington, Kentucky: University of Kentucky Press.

Miller, Russell. 1985. *The House that Getty Built.* New York: Henry Holt & Company.

Murphy, K.J. 1985. "Corporate performance and managerial remuneration: An empirical analysis." *Journal of Accounting and Economics* 7: 11-42.

Nation's Business. 1968. "Organizing for growth." August, 60-65

Neff, Thomas. 1984. *The International Uranium Market.* Cambridge, Massachusetts.: Ballinger Publishing Company.

Norman, James, Scott Ticer, and William Glasfalt. 1985. "ARCO enters oil's new era." *Businessweek.* May 13, 24-25.

Nulty, Peter. 1983. "Boone Pickens, company hunter." *Fortune.* December 26, 54-60.

Norman, James, Mary Pitzer, and Elizabeth Ehrlich. 1985. "The sharks keep circling Phillips." *Businessweek.* February 11, 24-5.

Penrose, Edith T. 1959. *The Theory of the Growth of the Firm.* Oxford, England: Blackwell.
Penrose, Edith T. 1968. *The Large International Firm in Developing Countries.* London: George Allen and Unwin Ltd.

Pierson, Walter R. 1976. *Witness for Oil.* Patricia Malony Markun (ed). Washington, D.C.: American Petroleum Institute. 936-49.

Rosen, S. 1982. "Authority, control, and distribution of earnings." *Bell Journal of Economics* 13 (Autumn): 311-23.

Rubin, P. 1973. "The expansion of firms." *Journal of Political Economy* 81   (July/August).

Salter, Malcolm S., and Wolf A. Wenhold. 1979. *Diversification Through Acquisition: Strategies for Creating Economic Value.* New York: The Free Press.

Scott, Otto J. 1968. *The Exception: The Story of Ashland Oil and Refining Company.* New York:McGraw-Hill Book Company.

Singh, Harbir, and Cynthia Montgomery. 1987. "Corporate acquisition strategies and economic performance." *Strategic Management Journal* 8: 377-386.

Stuart, Alexander. 1981. "What makes Mobil run?" *Fortune* December 14, 93-9.

Teece, David J. 1982. "Towards an Economic Theory of the Multiproduct Firm." *Journal of Economic Behavior and Organization* 3 (March): 39-63.

Thompson, Craig. 1951. *Since Spindletop: A Human Story of Gulf's First Half-Century*. Pittsburgh: Gulf Oil Company.

Vogel, Todd. 1989. "Phillips climbs up from the bottom of the barrel." *Businessweek*. January 16, 76-7.

Weinberg, Steve. 1989. *Armand Hammer: The Untold Story*. Boston: Little, Brown, and Company.

Welty, Earl, and Frank Taylor. 1966. *The 76 Bonanza*. Menlo Park, California: Lane Magazine and Book Company.

Williamson, O. E. 1967. "Hierarchical control and optimum firm size." *Journal of Political Economy* 75 (April) 123-38.

Williamson, O.E. 1981. "The modern corporation: Origins, evolution, attributes." *Journal of Economic Literature* 10 (December): 1537-68.

Williamson, O. E., M. Wachter, and J. Harris. 1975. "Understanding the employment relation: The analysis of idiosyncratic exchange." *Bell Journal of Economics* 6 (Spring): 250-80.

Index